UNDERSTANDING ELECTRONIC CIRCUITS

UNDERSTANDING
ELECTRONIC CIRCUITS

IAN R. SINCLAIR
B.Sc., MIEE, A.Inst.P

Fountain Press London

Fountain Press,
Model and Allied Publications Limited,
Book Division,
Station Road,
Kings Langley,
Hertfordshire, England.

First Published 1973

ISBN 0 85242 285 7

Printed in Great Britain by Offset Litho at
Gilmour & Dean Ltd., Hamilton, Scotland.

Bound in Great Britain at
Mansells (Bookbinders) Ltd., Witham, Essex.

CONTENTS

PREFACE

LIKE ITS COMPANION VOLUME, *Understanding Electronic Components*, this book deals with one section of the vast subject of electronics, in this case the basic types of circuits which are the daily concern of the electronics engineer. Whereas the previous book concentrated on the components, this book deals with the circuits in which the components operate. Inevitably, some previous knowledge is assumed; in this case of the properties of electronic components and the elementary laws of electrical circuits.

From this starting position, the book explains amplification, oscillation, switching and logic circuits, with an emphasis on understanding the principles involved rather than on rigorous theory. Particular attention is paid to the rapidly growing use of integrated circuits in electronics, and the merits and limitations of such circuits are discussed in connection with each different application.

Throughout the book, illustrated examples are taken from modern circuit practice, and these illustrations are by no means confined to the familiar consumer products, but also represent industrial, military and communications equipment.

This is not a book for the absolute beginner, but for those who have some circuit wiring experience but who may have little idea how the circuits work. As such it should be of considerable help to the amateur seeking to extend his knowledge, especially of circuits in equipment normally inaccessible to him, to the technician engineer approaching electronic circuitry with the aim of working on industrial or communications equipment, and to the service engineer looking for the principles of circuit action and guidance on possible faults.

This book should also have considerable appeal to students following a modern 'A' level Physics course which includes Electronics among the subject matter, and to the undergraduate seeking to obtain an overall picture of electronic circuitry in connection with project or similar work.

A book such as this can be produced only with the wholehearted co-operation of the electronics industry, and I wish to thank those who gave up their valuable time to discuss circuits and applications with me. I also wish to acknowledge the help, information and other assistance unstintingly given by the many firms approached during the writing of this book, in particular, the following:

Acoustical Manufacturing Co., Advance Electronics Ltd., A.P.T. Ltd., Brandenburg Ltd., British Radio Corporation, Electrocomponents (Radiospares) Ltd., Farnell-Tandberg Ltd., Ferrograph Co. Ltd., Jermyn Industries, H. J. Leak, Motorola Semiconductors Inc., Mullard Ltd., Philips Ltd., Pye Telecommunications Ltd., GDS (Sales) Ltd. (distributors of Motorola semiconductors), Tandberg Radiofabrikk A.S., Telequipment Ltd., Texas Instruments Ltd.

I must also mention the kind permission of the Editor of *Mullard Technical Communications* in allowing me to reproduce some circuit information from that most helpful journal. Finally, I must thank my wife for her tolerance, understanding and sheer hard work during the preparation of this book.

I. R. Sinclair *July, 1972*

CHAPTER ONE

SUPPLIES AND BIAS

AN ELECTRONIC SIGNAL IS A CHANGE, or series of changes, of voltage or current in a circuit. The signal may start off as an electronic signal (as, for example, the signal picked up by a radio) or it may be the result of the conversion of a mechanical signal (as in a record pickup) or a light signal (as in a TV camera) or various other types of signal. Strictly speaking, a circuit is any conducting path in which current can flow, but in electronics we usually use the word to mean the more complicated arrangements which alter the signal in some way, such as increasing amplitude (amplification), counting signals, shaping signals, guiding different signals to different routes.

Active and Passive Components

An electronic circuit can include active or passive components and most circuits contain both types. Passive components cannot increase the power of a signal; in fact, the power of a signal passing through passive components is always reduced. Active components use the signal to *control* another voltage or current, so creating another signal which can be used in place of the first one. This signal may be greater in amplitude, in which case we say that the active component has amplified the signal, though in fact it has created a new signal of greater amplitude under the control of the original signal.

If the new signal is, in every way, a 'scaled-up' version of the old one, we say that the amplification is *linear*; if not, we say that the amplifier is non-linear, the output is distorted. In some cases we are not interested in the shape of the signal which is created in the active device, but we may be interested in its size, the time at which it occurs, or some other points about the signal.

The main difference which distinguishes active from passive devices is the provision of power supplies. Passive components work on the signal alone; active devices need a supply of electrical energy in order to create the new signal. In addition, bias voltages or currents must be supplied to active devices so that the input signal can control the voltage or current at the output correctly.

Passive Circuits (D.C.)

Bias is supplied and adjusted by passive circuits, working with direct current. Capacitors act as insulators as far as d.c. is concerned, and inductors act as resistors. Note that this is not true for the first fraction of a second when a voltage is applied to a circuit. We are talking here of steady current flow conditions. The flow of currents in circuits consisting of steady voltage supplies and resistors is calculated by using Ohm's Law and the two Kirchoff Laws.

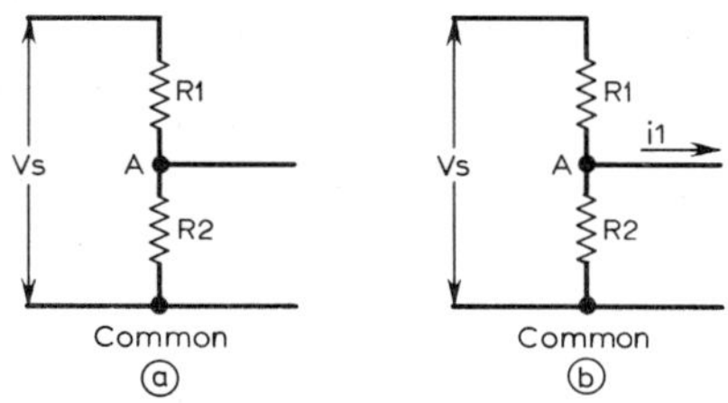

Fig. 1.1 (a) Potential divider circuit. (b) Current taken from divider circuit.

Where a steady voltage at less than supply voltage is needed, the potential divider circuit of Fig. 1.1 is often used. The voltage at the point A is

$$V_s \frac{R2}{R1+R2}$$

(measuring the voltage from point A to 'earth' or 'common'). This is no longer true if current is taken from point A to another circuit, when the voltage at point A becomes

$$\frac{R2}{R1+R2} \times (V_s - i_1 R_1)$$

If the current i_1 changes, then the change of voltage at point A is less if $R1$ and $R2$ are small resistors than if they are large. On the other hand, if no current is taken from A, the values of $R1$ and $R2$ may be as high as is convenient to use.

The point to which the bias is supplied is usually the point at which a signal enters an active device, and it is sometimes necessary to prevent this signal entering the bias supply, sometimes because the bias supply has so low a resistance as to reduce the signal strength, or because one bias supply feeds two active devices. In such cases, a resistor is connected between the potential divider and the point (B) to be biased, and a capacitor is connected between point A and earth.

The capacitance value should be such that its reactance for the lowest signal frequencies is much less (about ten times less) than the resistance of the potential divider from A to earth. This value of resistance is usually $R1 \times R2/(R1 + R2)$ because the power supply is connected by low-reactance capacitors to earth, so making $R1$ a connection to earth *as far as signal frequencies are concerned.* For this reason, $R1$ appears to be in parallel with $R2$ for signal frequencies. When this has been done, the potential divider is said to be 'decoupled'.

Bias supplied to the base of a transistor, the gate of a MOSFET or the grid of a valve, is of the type described, but the inputs to integrated circuits may need different arrangements. Where linear ICs are used, a very small change in bias may cause such a large change at the output as to cause the amplifier to become unusable. In such a case, a feedback bias circuit is arranged in which the bias supply is taken from the output of the IC, so that a change in the output causes a change in bias *but in an opposite direction.* For example, the circuit might be wired so that a rise in voltage at the output causes a rise in bias at an input terminal where a rise in input causes a *fall* of output voltage. This type of circuit, a feedback bias circuit, is self-correcting once the correct values for bias have been chosen.

Taking the typical arrangements applying to an operational amplifier (a high-gain amplifier), the usual conditions of operation require equal positive and negative power supplies so that both input and output are at about earth potential. In most cases, two input terminals are provided, one (−input) giving an inverted signal at the output and the other (+input) giving a signal of the same polarity at the output. If any connection is made between the output and the (−)input, this will stabilise the operating conditions, as a rise or fall of voltage at the output will be applied to the input to produce a correcting current within the circuit.

As any connection will accomplish bias, the resistor used to join output to (−)input is chosen so as to give the conditions of amplification and input resistance desired (see later). In these operating conditions, the (+)input of the 'op-amp' is earthed, either directly or through a resistor.

It is also possible to operate these IC amplifiers using a single voltage power supply. Since the output of the amplifier depends on the voltage difference between the two inputs, any voltage to which the two inputs are returned will cause the output to take up about the same voltage. If both inputs are returned to a voltage which is about half the supply voltage, and a connection is again made between the (−)input and the output terminal, the op-amp will automatically bias itself to the half-supply voltage.

Other terminals of active devices may be biased. Very often the emitter of a transistor, the source of a MOSFET or the cathode of a valve is biased, sometimes because bias at this point is easy to arrange or because a signal can then be applied or taken from this point. In each case, a resistor is connected between the active point and earth (or common) line and the bias voltage is the voltage developed across the resistor by the current flowing through it (Fig. 1.2). In MOSFETs (and valves) where a positive voltage on the source (or cathode) is equivalent to a negative bias voltage on gate (or grid) this is often the only type of bias used.

If the amplification of the device is not to be lowered by the use of a bias resistor in this position (because the signal current will cause a signal voltage) then the bias resistor must be decoupled by a capacitor in parallel. The value of this capacitor is given approximately for transistors by the formula $C = 6{,}000\, I_c/f_1$, where I_c is the collector current (no signal) in mA, f_1 is the frequency at which the amplification is 3dB down (in Hz); C is in μF. For valves and FETs the formula is

$$C = \frac{gm + 1/R}{2\pi f} \times 1{,}000$$

where gm = mutual conductance, R = bias resistor in kΩ, f_1 is frequency as before (in Hz). (These formulae are due to P. Engstron, Lund Institute of Technology Sweden, *see Wireless World, December, 1971.*)

The output terminal of an active device is not usually regarded as being biased, but the voltage at this point should be approximately half of the supply voltage if a linear amplifier is being used, and the calculations needed form part of the calculation of biasing. The requirement for half-supply-voltage, together with the current fixed by the previous arrangements, decides what value of resistor will be used, and the choice of this resistor (the load resistor), is not free; it may be an external load (a loudspeaker, a tape head) or it may have to be a low value of resistor because of frequency limitations (see later).

When the active device is not used as a linear amplifier, biasing

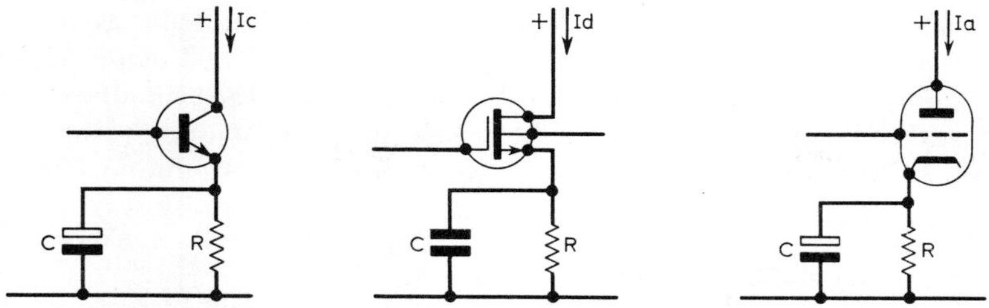

Fig. 1.2 Use of biasing resistor in emitter, source or cathode circuit. (Other connections omitted for clarity)

is much easier. The bias circuit will be arranged so that the active device either passes *no* current, or passes such a high current that the output voltage (across the load resistor) can go no lower. The device is 'cut-off' or 'saturated'. In the cut-off condition, with no current normally passing, only signals which cause current to pass can produce an output; conversely, in the saturated condition, only signals which reduce the current produce an output. Such signals are pulse or digital signals.

Passive Circuits (A.C.)

In *resistors*, alternating currents behave exactly as direct currents, providing that the resistors are behaving purely as resistors. This is true whether the a.c. is in the form of a sinewave, a pulse or some other waveform. If we examine the waveforms of the alternating voltage and the alternating current, we find that they occur at the same time and they are said to be in phase.

In *capacitors*, the waveform of a sinewave of *current* is ahead in time of the wave of voltage; the current and voltage are out of phase. In this case, where current *leads* voltage, the amount is $\frac{1}{4}$-wave, or 90°. In addition, when current passes through a pure capacitance (no resistance or inductance present) there is no power dissipated to heat the capacitor. When pulse waveforms are applied to a capacitor, the voltage and current waveforms are not separated in time (unless the pulses repeat rapidly, forming a continuous waveform) but are entirely different in shape.

In *inductors*, we encounter similar effects. With a sinewave, the inductor has a voltage across it which leads the current through it by 90°, if the inductor is resistance-free, but by less than 90° in practical cases. Pulses cause different waveforms of voltage and current with a specially important case that the sudden stopping of current in an inductor causes a large rise of voltage across the inductor in a direction which tends to keep the current flowing. These points, and others concerning the behaviour of resistors, inductors and capacitors, are described more fully in the companion volume to this book, *Understanding Electronic Components* (Fountain Press).

Metering and Measuring (D.C.)

Current measurements on d.c. are made with moving coil meters almost exclusively. The current to be measured (or a known fraction of it) passes through a coil of fine wire which lies between the pole-pieces of a powerful magnet whose flux (lines of force) is directed towards the axis round which the coil can move. When a current is passed through the coil, the magnetic flux acts on the moving charges in the wire (current) to give a twisting force (torque), so turning the coil on its axis. This is resisted by a spring (a fibre in very sensitive instruments) so that a given value of current causes a definite deflection. If no springs were used, any value of current which could move the coil would twist it round to the position where the gap between the poles interrupted the flux.

Only current in one direction can turn the coil in the correct direction, and a stop, bearing usually against the indicating pointer and fastened to the coil (or its axle), prevents movement in the opposite direction in most instruments. Some moving coil instruments are arranged to have a centre-zero, and can record current in either direction, and some replace the pointer with a mirror, so that a light-ray (which has no weight and no inertia) can be used as an indicator.

Because of the resistance of the coil, there will always be a voltage drop across a meter when current passes through. The size of this voltage drop at full scale deflection can be found by multiplying the meter resistance (the resistance of the moving coil, disregarding shunts) by the current for full scale deflection (again, disregarding shunts).

For example, a 1mA, 100Ω meter will have a voltage of 100mV across it at full scale deflection, and this will be true no matter how the meter is shunted for other ranges of current. In many cases, this voltage drop is of no importance; in some circuits, however, which operate with small values of voltage and current bias, this may cause the operating conditions to change noticably when the meter is inserted in circuit. In such cases, meters with very small operating voltages should be used.

The voltage drop across a meter movement enables us easily to protect meter movements against overloads. If two diodes are wired across the movement so as to allow current to flow in either direction when their forward voltages are exceeded, then the diodes will conduct when the voltage drop across the meter movement (in either direction) is greater than the forward voltage of the diodes.

Because of the shape of the current/voltage characteristics of a diode, the share of the overload current taken by the diodes becomes very considerable when the voltage across the movement is only slightly greater than the forward voltage of the diodes, so protecting the meter from large overloads. The advantage of using diodes as compared to the earlier electro-magnetic trips is that diodes respond to a short period overload.

Current Ranges

Any basic moving coil meter can be adapted to read larger full-scale values of current by connecting a *shunt*, a resistor in parallel

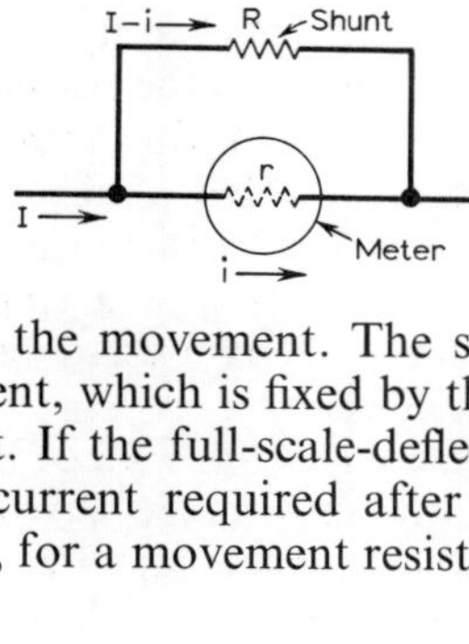

Fig. 1.3 Use of shunt with meter movement.

with the movement. The shunt (Fig. 1.3) takes a fraction of the current, which is fixed by the resistance values of shunt and movement. If the full-scale-deflection current of the movement is i and the current required after shunting for full-scale-deflection is I, then, for a movement resistance of r:

$$R = \frac{ir}{(I-i)}$$

where R is the shunt required. If I and i are in the same current units (Amps or mA etc.), then R and r will be in the same resistance units (ohms or kΩ).

Voltage Ranges

The same type of moving coil movement is used in voltmeters, but with a resistor, or a set of switched resistors, in series with the moving coil. The voltage measured at full-scale is then given, using Ohm's Law, by $V = R \times i$ where R is the total resistance (including

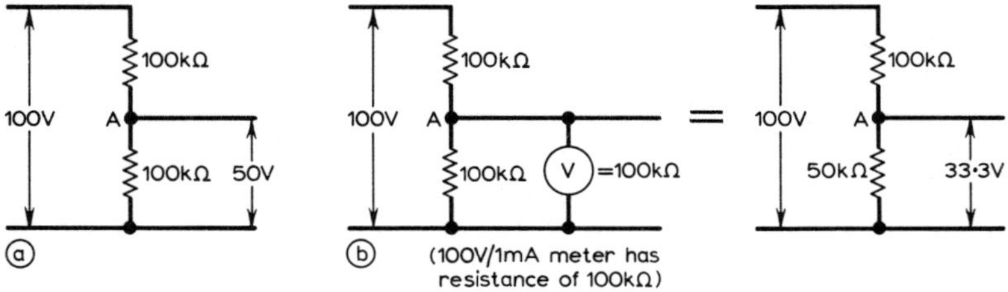

Fig. 1.4 Loading a potential divider by a meter: (a) No load, output voltage 50V; (b) Meter acts as 100kΩ load in parallel with 100kΩ resistor. The effect is to reduce the total resistance from A to earth to 50kΩ, so reducing the voltage to 33·3V.

the moving coil) and *i* is the full-scale-deflection current. In this case it is the current taken by the meter which can prove troublesome, as it has to be supplied by the network feeding the point whose resistance is being measured. Fig. 1.4 illustrates the principle; the voltage at A is 50V when no meter is present, but when a 100V meter taking 1mA f.s.d. is used, the reading is 33·3V.

A good general rule when using voltmeters is not to use an instrument whose resistance is less than ten times the resistance in the circuit which will supply current to it. For this reason, the ohms-per-volt factor of a meter is often quoted. This figure is simply 1/f.s.d.; its use is to calculate quickly the resistance used in the meter. For example, if a 20kΩ/V meter is used on a 10V range the resistance used is 200kΩ, and this can be safely used to measure, for example, a bias chain fed by a 10kΩ resistor.

Where a high input impedance transistor or FET stage is used, it is a waste of time trying to use a low impedance moving coil meter to measure bias conditions. If no other instrument is obtainable, the bias conditions must be deduced from the voltage at the collector (or drain); alternatively a valve or FET voltmeter with a very high input resistance can be used. When the bias conditions

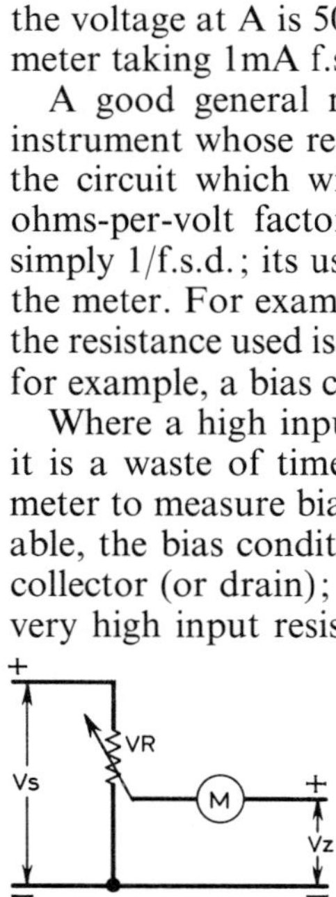

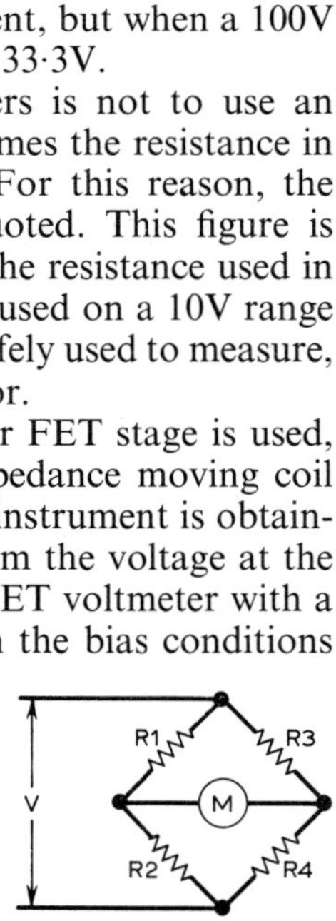

Fig. 1.5 (left) The Potentiometer. The variable resistor delivers a voltage Vs·X, where X is the division ratio. If this equals Vz, the meter M is undeflected. The value of X can be read off the dial of the variable, and X·Vs calculated.

Fig. 1.6 (right) Wheatstone Bridge. R1, R2 is usually a tapped variable with the ratio read from a dial.

of integrated op-amps, or groups of transistors using feedback bias, are to be checked, it is always better to check the voltages at the collector and emitter points where available.

Small changes in the voltage of the feedback loop, caused by the meter, can cause very large changes in the bias conditions. This is particularly true of integrated op-amps with very high gains; no attempt should be made to measure the input bias voltages.

Potentiometers and Bridges

The name 'potentiometer' has unfortunately come to be used for a three terminal variable resistance; in the original sense it is a very sensitive method of voltage measurement *which takes no current from the circuit being measured.* Because it must be calibrated and is not a simple meter reading it has been neglected in electronics use, but is still a very useful method of measuring 'difficult' voltages. The principle of the potentiometer is shown in Fig. 1.5.

A known voltage source, which may be a mercury cell for low voltages or a stabilised supply for high voltages, is connected across a tapped resistor whose tap ratio is accurately known. This may be a good linear potentiometer or a helipot with a digital dial reading 0 to 1000. The centre tap is connected to a current detector (a moving coil galvanometer or a milliammeter connected to an op-amp) and so to the unknown voltage through a key-switch.

When the key is tapped, current will flow from the higher voltage to the lower; but if the voltage supplied from the tapping point is exactly equal to the unknown voltage no current will flow and the detector will indicate this. When this is the case, there is obviously no current taken from the source of the voltage being measured.

The Wheatstone Bridge is an old circuit which is still much used in resistance measurements. It consists (Fig. 1.6) of a network of four resistors, of which two are formed by the arms of a variable resistor, one is a resistor of accurately known value and one is the resistor to be measured. When no current can be detected by the current meter, $R1/R2 = R3/R4$. The ratio $R1/R2$ can usually be read directly from the dial attached to the variable resistor.

A bridge of this type is only as accurate as its fixed resistor, so that wirewound resistors of good temperature stability should be used for the lower ranges of resistance and metal film resistors for the higher values. Balance values obtained near the end of the scale on the variable resistor are unreliable, and the ranges in a multi-range instrument should be arranged so that it is never necessary to use the extremes of the dial.

Metering and Measuring (A.C.)

The metering and measuring of a.c. presents more difficulties, as frequency and waveshape must be taken into account as well as voltage or current. At mains frequencies of 50Hz, moving iron instruments can be used. These work on the principle of using current in a fixed coil to induce magnetism in a piece of iron which is mounted on pivots, carries a needle pointer and is controlled by springs in the same way as the coil of the moving coil instrument.

In action, the iron has induced in it a magnetic field equal and opposite to the field in the coil and so is attracted to the coil. This action is unchanged if the direction of the current in the coil is reversed, so that moving iron instruments can be used equally well on a.c. as on d.c. In addition, the quantity measured by a moving iron instrument used for a.c. is the r.m.s. value; the value of d.c. current or voltage which would have the same effect of providing power as the a.c. current or voltage.

The moving iron instrument is of more use in the 'heavy-current' side of electrical engineering. In electronics, peak voltages are of more interest and the non-linear scale of the moving iron meter and its very limited frequency response confine it to use at mains voltages and currents.

Much more use is made of moving coil meters fitted with rectifiers. The scale is linear and the useful frequency range greater, but the scales are usually 'arranged' to read r.m.s. values, despite the fact that the output of the rectifier measures the peak value. The scale assumes that a sinewave is being measured, as the r.m.s. value of a sinewave of V volts peak is $V/1{\cdot}414$.

This translation from peak value to r.m.s. is true only for a sinewave, so that the readings obtained on such a meter for any other waveform are inaccurate, sometimes ludicrously so.

Where waveforms which are not sinewaves are to be measured, the oscilloscope is the instrument universally used for this purpose. The measuring capabilities of oscilloscopes have improved over the last ten years to an extent where they can rival the cheaper moving coil meters in accuracy, yet provide measurements on waveforms which are otherwise unmeasurable. So important is the use of the oscilloscope that in this book it has been given a chapter to itself (Chapter 10).

This does not mean, however, that meter readings have no place in a.c. measurements. The power output of an audio amplifier, for example, will be taken using a sinewave input and measuring the r.m.s. voltage developed across the output load. In addition, to this, more specialised instruments such as distortion meters (which filter out and measure the amplitudes of frequencies which are multiples of the frequency fed in) may be used in audio work. Most of the instruments used, however, are adaptations of moving coil/rectifier meters.

The *field strength meter* is used to measure the strength of a radio frequency field at any place. It consists of an aerial, a circuit tuned to the frequency to be measured, a diode detector and a moving coil meter. In use, the frequency is picked up on the aerial, whose length is known, selected by the tuned circuit, which also ensures that a sinewave is passed to the diode, rectified and measured. The meter reading is usually in terms of μV/m, measuring the number of microvolts of signal at the frequency chosen per metre of aerial length used.

Field strength meters are of particular value in siting aerials and in plotting the effective coverage of a transmitter. Grid dip meters are used to measure the frequency to which a circuit is tuned, and are useful for checking the action of tuned amplifiers without actually applying power to the amplifier. They consist of an oscillator which can be tuned over a range of frequencies indicated on a dial. The output of the oscillator can be coupled to any tuned circuit by a loop of wire, and, when the oscillator is tuned to the same frequency, power is transferred from the oscillator to the tuned circuit.

When this happens, the current taken by the oscillator changes and the change, which takes place very sharply at the resonant frequency, is detected on a moving coil meter. In the original design of the instrument, a valve oscillator was used and the meter was placed in the grid circuit, showing a dip in the grid current at resonance, hence the name.

SWR meters are used to check the matching of transmission lines and waveguides and consist of diode rectifier and moving coil meters coupled to adjustable pieces of line or waveguide. An incorrectly terminated line causes reflections of waves which then set up a pattern of high and low voltage along the line. The rectifier and meter detect such peaks.

Frequency meters, apart from the vibrating reed type used for checking mains frequency, count the number of waves passing through a gate in an interval of time which is accurately controlled by a crystal oscillator. The frequency is then displayed on counter tubes. In the usual form of display, the measured frequency is displayed for a set time (usually 1 second) before being measured and displayed again, so that continual small changes of frequency do not cause reading difficulties. The circuits used will be examined more closely when we deal with counting circuits in a later chapter.

Measurements on Capacitors and Inductors

Capacitance values can be measured using a direct-reading capacitance meter. Referring to Fig. 1.7, a capacitor is charged up to a set voltage V and then discharged through a moving coil current meter. At each charge, the amount of charge on the capacitor plates is given by $C \times V$ (C in farads, V in volts), and when the switch changes over, this charge flows through the meter. As current is rate of flow of charge, if the switch changes over n times per second there must be a current of nCV amps flowing. If we measure C in μF, then the current nCV is in μA.

Typically the switching might be carried out by a reed switch at the mains frequency, giving two change-overs in each cycle (one on each voltage peak). Using a 100V supply, the current flow from a 1μF capacitor would be $100 \times 1 \times 100\mu$A, which is 10mA. A 1nF capacitor would give a reading of 10μA. The sensitivity can be improved by raising the voltage, raising the frequency of operation or using a more sensitive detector. Of these possibilities, the first can seldom be used, as so many capacitors designed for transistor circuits have working voltages well below 100V. The easiest way of improving sensitivity is to use an amplifier between the discharge path and the meter.

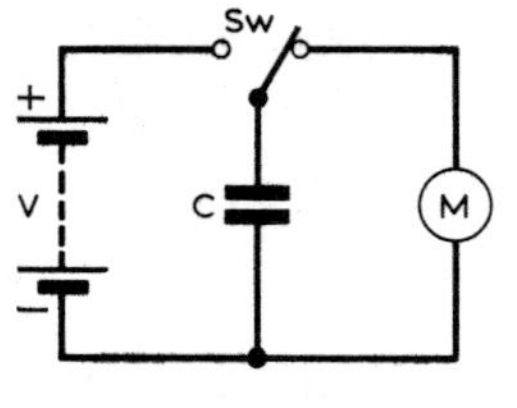

Fig. 1.7 Principle of direct-reading capacitance meter. The switch must operate continuously at a rate faster than the meter can follow.

This type of capacitance meter is simple, direct reading, and reliable, though it should not be used for large capacitance electrolytics where the charging time needed may not be available, and the discharge current can be large enough to damage the meter or amplifier.

Capacitance and Inductance Bridges

Both capacitors and inductors can be measured in bridge circuits similar to the bridges used for resistance measurements, but with modifications to allow for power factor. In most cases, the power factor of modern capacitors is so close to 1 that it is a waste of time to attempt to measure it unless with very sensitive transformer-arm

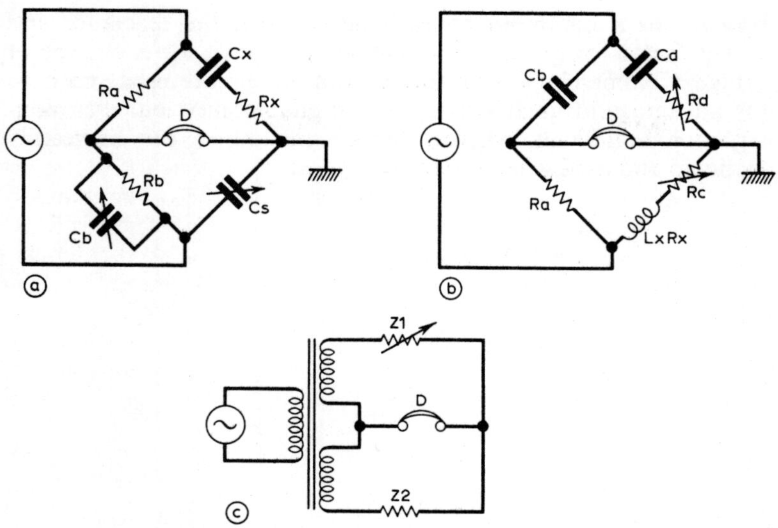

Fig. 1.8 Reactance Bridges. In each case, a.c. is applied and D is a sensitive detector. (a) Schering Bridge. At balance Cx = CsRb/Ra and Qx = 1/(CbRb). For most measurements, Cb can be omitted. (b) Owen Bridge. At balance Lx = CbRaRd, Rx = (CbRa/Cd) − Rc where Rx is the resistance of the coil L. (c) Simple transformer-arm bridge. At balance, Z1 = Z2 where the Z's can be any type of impedance. The transformer secondaries must be accurately matched. In other versions, the transformer tap is variable and used to obtain balance.

bridges, but the power factor of inductors is well within the capabilities of a simple bridge.

Bridges intended for reactance measurement require an a.c. supply. If only the larger sizes of capacitors or inductors are to be measured 50Hz mains can be used, but for measurements of μH or pF values a high frequency signal, internally generated, is more useful. It is important for bridge use that this signal should be a pure sinewave, otherwise it is difficult to obtain a clear zero reading (or minimum reading) on the detector.

As the detector has only to indicate a zero or minimum and need not be calibrated, any form of a.c. meter or display, from headphones upwards, can be used, and amplification is easily arranged. As usual with bridge circuits, variation of the supply voltage has no effect on the bridge balance.

Typical bridge circuits for capacitors and for inductors are shown in Fig. 1.8. In each case, two controls are used to bring the bridge to balance, though for capacitors, as explained, the power factor control seldom needs to be altered. When the detector is brought to a minimum by each adjustment, the reactance and power factor (or Q) can be calculated as shown for each type of bridge. Commercially, transformer-arm bridges are used and have the advantage of great sensitivity and adaptability, but their construction is difficult, whereas the circuits shown can be readily made up and used when needed.

CHAPTER TWO

AMPLIFICATION

IN THE EARLY DAYS OF ELECTRONICS when valves were the only known amplifying devices the subject of amplification hinged on the working principles of valves. It was not until transistors and other amplifying devices came into use that designers began to appreciate that the important points about amplification have nothing to do with the actual device used to amplify. For most purposes, the amplifier can be thought of as a 'black box' which is wired into a circuit and carries out certain operations to the signal.

The coming of integrated circuits has made this idea a reality; an amplifier unit (operational amplifier) can now be bought which performs according to a specification although none but the maker knows, nor cares, what lies inside. We shall discuss amplification throughout this chapter, therefore, without ever referring to a transistor or a valve, since the ideas of amplification do not depend on such devices alone.

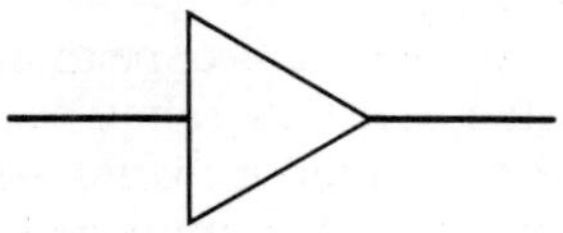

Fig. 2.1 Amplifier symbol. There may be more than one input or output, and other connections (internal) may be shown on the sloping sides.

The symbol for such an amplifier is shown in Fig. 2.1; it takes the form of an arrowhead pointing in the direction in which amplification takes place. In our example, the signal is fed in at the left hand side and emerges amplified at the right hand side. Inside the arrowhead, or below it, we may write down figures to describe the performance of the amplifier.

Gain

The gain of an amplifier is the amount by which it amplifies a signal; this may be expressed in several ways. The *voltage gain* is the ratio

V out/V in and can be measured readily with an oscilloscope. Both voltages must be measured in the same way; if the input signal is measured peak-to-peak, then the output must be measured peak-to-peak, and so on. For most purposes the voltage gain is by far the most useful figure of gain since it can be so easily checked using the oscilloscope (in the case of d.c. amplifiers, a meter is a more useful check). In the amplifier symbol of Fig. 2.1 the voltage gain may appear as $G = -20$ or $G = +20$. In each case, this means that the p-p voltage of the output is twenty times that of the input, but the − or + signs also have a meaning. The − sign indicates that the amplifier is *inverting*, the signal at the output is upside down compared with the input; the + sign indicates that the amplifier is *non-inverting*, the signal at the output is the same way up as that at the input.

This is shown in Fig. 2.2 for a waveform of square pulses and also for a sinewave. The same waveforms are also shown shifted in phase by 180° ($\frac{1}{2}$-cycle) and it can be seen that the inverted sinewave is the same as a sinewave phase-shifted by 180°, but the inverted pulse waveform is not the same as the phase-shifted version. A waveform which looks the same inverted as it does when phase-shifted 180° is called a symmetrical waveform. In some texts, inversion and 180° phase-shift are treated as identical; this is true *only for symmetrical waveforms*, as can be seen above.

Power gain is sometimes quoted. It is important in the case of power amplifiers feeding signals to loudspeakers, relays, aerials or any other device which consumes electrical power. Power gain is: signal power output of amplifier/signal power at input to amplifier, and is the same as voltage gain × current gain.

Current gain may be measured, using the oscilloscope to measure the signal voltages across small equal resistors (about 1Ω) one placed in series with the input and the other in series with the output. Since the resistors are small, the currents at input and output are not affected, and the voltages across the resistors are equal to resistance times current. Since the resistances are equal, the ratio of voltages must be equal to current gain. In many cases, as will be seen later, current gain can be calculated, but for many purposes of servicing work, power gain does not have to be known.

Decibels

For many purposes, gain is measured in decibels (dB), a unit which the electrical engineer has borrowed from the acoustical engineer. Voltage gain expressed in decibels is 20 log (Vout/Vin); for example,

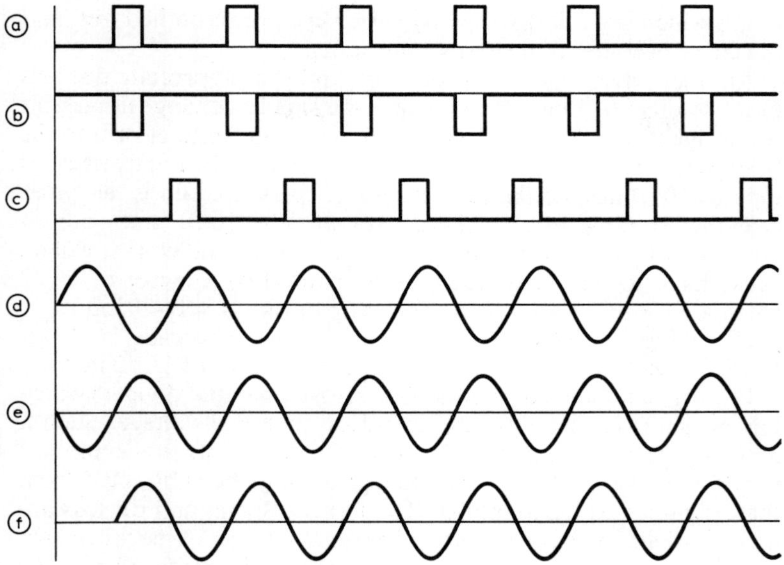

Fig. 2.2 Inversion and phase shift. (a) square pulses, (b) inverted square pulses, (c) original square pulses phase shifted by 180°, (d) sinewave, (e) inverted sinewave, (f) original sinewave phase shifted by 180°.

if an amplifier has a gain of 10 (log 10 = 1), the gain in dB is 20dB; if the amplifier gain were 100 (log 100 = 2), the gain in dB would be 40dB; if the amplifier gain were 1,000 (log 1,000 = 3), the gain in dB would be 60dB. This notation has several advantages. One is that decibels can be added, whereas gains must be multiplied.

If a signal is fed through a network with a loss of 10 times into an amplifier with a gain of 100 times and out through a network with a loss of 5 times, the gain is given by $100 \times 1/10 \times 1/5 = 2$. In dB this becomes a gain of 20 log 100 (40dB) and losses of 20dB and 14dB, a total of 6dB, which agrees with the gain of 2. If each stage in an amplifier is responsible for a gain or loss expressed in dB, then the total calculation of gain amounts only to adding and subtracting the decibel gains and losses.

In many cases, gain and loss is only ever expressed in dB terms; for example, the graph of gain against frequency uses dB units for gain and plots against frequency on a logarithmic scale in which equal distances along the scale represent tenfold increases in frequency. On such a graph, small irregularities in the gain/frequency graph which are not detected by ear or eye are smoothed out, and the curve becomes more easily interpreted.

Another advantage is that our ears and eyes appreciate decibels more readily; 1dB increase in gain is the smallest change noticeable when signals are converted into sound by a loudspeaker or into the brightness of a cathode ray tube display, and 40dB sounds twice as loud as 20dB despite the fact that the increase in gain is ten times (see the example above where gain of 10 = 20dB and gain of 100 = 40dB). A third advantage is in *CR* or *LR* networks, where, as we have seen already, the gain is reduced by a factor of $1/\sqrt{2}$ when $X_c = R$ or when $X_L = R$. This drop in gain is 3dB ($20 \log 1/\sqrt{2} = 20 \times \frac{1}{2} \log 2 = 3{\cdot}010$) as near as makes no difference, and this figure is much easier to remember than the awkward $1/\sqrt{2}$ factor.

Fig. 1.1 has shown a network (a resistive potential divider) which has no gain but actually causes a drop in signal voltage. Such a network, properly called an *attenuator*, can have its loss expressed in dB, and the dB lost in the attenuator must be subtracted from the dB gained in the amplifier. The terms dB gain and dB loss are used to distinguish *amplification* from *attenuation*. A negative sign can also be used for attenuation, this must not be confused with the negative sign used to mean an inverting amplifier.

Where *power* gains are being compared, the formula for calculating decibels is changed to 10 log (Pout/Pin). For this, and the reasons above, the engineer prefers to work with voltage gains alone. The use of the 20 log (Vout/Vin) where volts (or amps) are measured and 10 log (Pout/Pin) where watts are measured is rather confusing at first sight. The original definition of the decibel was as one tenth of a larger unit, the bel, which was defined as log (Pout/Pin) for a network. The bel is rather a large unit, and the decibel, one tenth of the amount, so that 10dB equals one bel, is universally used. Note that the definition is for power only.

If a network has the same input and output resistance, R, then the power input is V^2 in/R and the output power is V^2 out/R, where V_{in} and V_{out} are the voltages across the input resistance and the output resistance respectively. If we substitute these quantities into the definition of the decibel, we get

$$\text{dB} = 10 \log \frac{V^2\,\text{out}/R}{V^2\,\text{in}/R} = 10 \log \frac{V^2\,\text{out}}{V^2\,\text{in}} = 20 \log \frac{V\,\text{out}}{V\,\text{in}}$$

(since $\log V^2 = 2 \log V$). This definition is equally valid *when the input and output resistances are equal*. In the same way, we can measure input and output currents producing power (in equal input and output resistances), $I_{in}{}^2R$ and $I_{out}{}^2R$; this gives dB = $20 \log I_{out}/I_{in}$.

These alternative definitions are *not* strictly true when the input and output resistances are different, as the powers are no longer simply proportional to the square of voltage or current, but the 20 log V_{out}/V_{in} notation is very widely used. Where the dB gain of the same amplifier is compared at different frequencies, or where the voltage or current gains are of more importance than power gain, the use of the 'voltage' decibel is justified, providing no attempt is made to compare these figures with decibels of genuine power gain.

Phase Shift

An amplifier affects the phase of a signal as well as its amplitude, for any signal must take some time to pass through an amplifier, and if the amplifier contains any networks whose time constants affect the signal, then this, too, must have some effect on the phase. The phase shift may be expressed as a fraction of a cycle or as a number of degrees, usually the latter.

Measurement of phase *shift* must be carried out using a sinewave signal since phase shift has the effect of changing the shape of other waveforms, making measurements (using the oscilloscope) difficult. Phase *delay*, caused by the time which signals take to pass through an amplifier, can be measured only by a first-class measuring oscilloscope, as delay time is usually very small. The signal used is a very short pulse.

The time delay is related to the phase shift of sinewaves. If we had an amplifier whose only phase effect on a pulse was to delay it, such an amplifier would shift the phase of a sinewave by an amount proportional to the frequency of the sinewave, so that low frequency signals would be only slightly shifted (measured by degrees of shift) and high frequency signals greatly shifted. This is because an equal shift in *time* is not an equal shift in *phase* for signals of different frequencies.

When a phase shift is due to a *CR* or *LR* time constant, the phase has shifted by 45° when the amplitude response is 3dB down (0·7 of its 'normal' value). A single *CR* or *LR* time constant cannot cause more than 90° of phase shift (and this only at frequencies where the amplitude is very low), but a combination of time constants may cause larger phase shifts. If a signal at the output of an amplifier which is in the same phase as the signal at the input is *fed back* (see later), oscillation starts and this may occur in an unplanned way due to unwanted phase shifts in an amplifier, either at higher or lower frequencies than the planned frequency limits.

Input and Output Impedances

If a signal is applied to the input of an amplifier, and oscilloscopes are suitably set up, both the voltage of the input signal and its current can be measured. Voltage and current are always related by Ohm's Law, $V = ZI$ (or $Z = V/I$), and so the amplifier input has an impedance Z_{in} given by the ratio: Input signal voltage/Input signal current. This impedance may act like a resistance (resistive impedance), like a capacitance (capacitive impedance) or like an inductance (inductive impedance); most usually it behaves like a resistance with a capacitance in parallel (complex impedance) with some definite time constant.

We can, in the same way, measure an impedance, the output impedance Z_{out}, at the output of an amplifier. These impedances affect the way in which we can couple an amplifier to other parts of a circuit. Fig. 2.3 shows our amplifier symbol with the input and output impedances drawn in separately (often we simply note under the symbol the values of Z(in) and Z(out)). Z(in) is between the amplifier input and earth; Z(out) is in series with the output of the amplifier, which is the simplest case usually encountered; also the impedances are shown as pure resistances.

Suppose, for example, the values of Z(in) and Z(out) are, as shown, 100kΩ and 20kΩ respectively, these impedances then act as potential dividers with any other resistances in circuit and Fig. 2.3 shows the effect of feeding the amplifier from a 100kΩ resistor and into a 20kΩ resistor. The 100kΩ resistor and the 100kΩ input impedance halve the input signal; the 20kΩ output impedance and the 20kΩ resistor also halve the output signal. In all, the signal has been reduced to $0{\cdot}5 \times 0{\cdot}5 = 0{\cdot}25$ of what it would have been if the impedances had no effect.

If the voltage gain of the amplifier were 20, the overall gain would be $0{\cdot}25 \times 20$, which is 5, because of the effect of the impedances. To obtain more nearly a gain of 20, the resistance in series with the input would have to be very small and the load

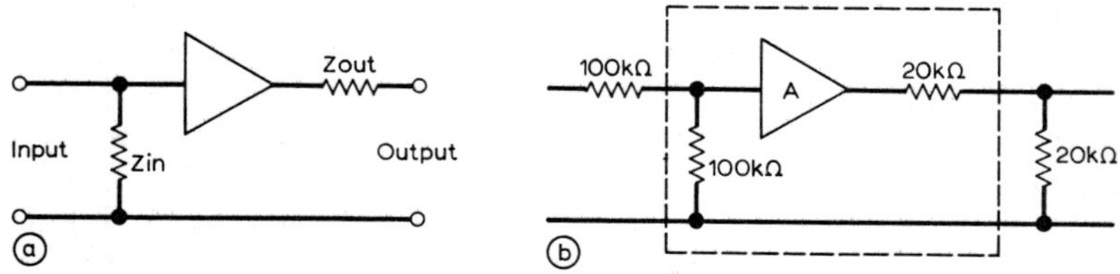

Fig. 2.3 (a) Amplifier with input and output impedances drawn in, (b) effect of source and load impedances on gain of a voltage amplifier.

resistance very high. In addition to the input and output impedances, the backward resistance (which contributes feedback) may also be quoted.

Amplifier Distortion

A perfect amplifier would amplify signals by the same amount regardless of their frequency or amplitude, and would leave their phase unchanged, or else change it in such a way that the output waveform was always the same shape as the input waveform.

There is no perfect amplifier, and for many purposes we can use amplifiers which fall a long way short of this degree of perfection. We call any departure from perfection in an amplifier *distortion*, and we divide distortion into two types: frequency distortion, in which an amplifier behaves well for only a limited range of frequencies, with unacceptable changes of gain and phase outside this limited range, and non-linear distortion, in which the gain of the amplifier varies with the amplitude of the signal at the input.

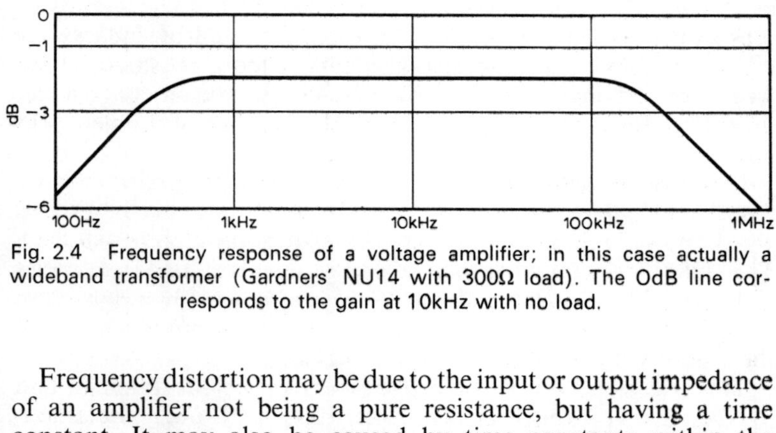

Fig. 2.4 Frequency response of a voltage amplifier; in this case actually a wideband transformer (Gardners' NU14 with 300Ω load). The 0dB line corresponds to the gain at 10kHz with no load.

Frequency distortion may be due to the input or output impedance of an amplifier not being a pure resistance, but having a time constant. It may also be caused by time constants within the amplifier or other effects (such as stored charge in transistors or transit time in valves) which cause the same effects as a time constant. For every amplifier, a graph of frequency response can be drawn showing how gain and phase vary with frequency. Fig. 2.4 shows a typical gain/frequency curve with gain in dB plotted against frequency in Hz. Note that the graph is drawn so that the frequency scale progresses in steps where equal distances represent a tenfold increase of frequency. This type of graph (logarithmic) is used to fit in a large range of frequency and also because the dB scale is logarithmic.

Frequency response may be checked in two different ways. If a quick check is wanted, the oscilloscope may be used to view the output of the amplifier when squarewaves of various wave-times are fed into the input. Integrating time constants, which cause loss of gain at high frequencies, lengthen the rise-time of a squarewave; differentiating time constants, which cause loss of gain at low frequencies, cause the flat tops of a squarewave to droop. The amounts of increase in rise time and percentage droop can be related to the frequencies where gain is 3dB down by Table 2.1.

The range of frequencies between the low frequency and the high frequency points where response is 3dB down is termed the bandwidth of the amplifier. The figure obtained by multiplying the gain of an amplifier by its bandwidth is called the *gain-bandwidth* (G-B) factor, and depends mainly on the type and number of amplifying devices used.

Alternatively, the gain of an amplifier can be measured by comparing input and output voltages with an oscilloscope for sinewave signals of different frequencies so that a gain/frequency graph can be drawn. This is a slower procedure, but does detect humps or dips in the response which are not easily detected by squarewave checks. In service, squarewave checking is more common. If the frequency response of an amplifier has changed, the service engineer must look for a change in a time constant, either in the networks outside the amplifier or in the amplifier itself.

It is also possible for an amplifier to have a perfectly good frequency response for signals of small amplitude, but be limited when large amplitude high frequency signals have to be handled. The ability of an amplifier to handle large signals at high frequencies is described by a quantity known as *slew rate*, usually measured in volts per microsecond, which measures the maximum rate at which the output voltage of an amplifier can change.

This is determined by the capacitances within the amplifier and by the currents which are available (passing through the amplifying devices) to charge them. If a capacitance C (measured in Farads) is charged and discharged by a current whose maximum value is I (in amps), then the slew rate in volts/second is I/C. In the more common units of pF, mA and V/μS, this becomes $I'/C' \div 1{,}000$, so that if a current of 10mA is available to charge a capacitance of 5pF, the maximum slew rate is $10/5 \div 1{,}000$ which is 0·0020V/μS.

Non-linear distortion can have several causes. One of the causes is that most amplifying devices normally conduct in only one

(a) Integrating time constants and sag.

If t is the time taken for the top of a square pulse to sag by 10%, f is the frequency at which the amplifier may be expected to be 3dB down.

t	f
1·0mS	166Hz
10mS	16·6Hz
100mS	1.66Hz

(b) Differentiating time constants and rise time.

If t is the time taken for the side of a square wave to rise (or fall) to 90% of its final level, f is the frequency at which the amplifier may be expected to be 3dB down.

t	f
10μS	37kHz
1μS	370kHz
100nS	3·7MHz
10nS	37MHz

Table 2.1

direction, and must be arranged so that some current is always flowing (biasing). If the input signal is large enough to cause the current to try to reverse (which it cannot do), that part of the input signal greater than the amount needed to stop current flowing is not amplified, causing distortion due to overloading at the input.

Another form of distortion can take place at the output of an amplifier. Since any amplifier must use a power supply, the peak-to-peak amplitude of voltages inside the amplifier cannot be greater than the amplitude of the supply voltage. For example, if the supply is 12V, then 12V is the maximum peak-to-peak amplitude of signal which can be handled, and signals of excessive voltage are cut off, a process called clipping.

The voltages at which clipping occurs may be deceptive. If an amplifier has a 3:1 step-up transformer at the output, a 36V clipping level might occur in an amplifier which clipped at 12V (the apparently better signal at the output being the smaller clipped signal stepped up). In such cases, measurements would have to be carried out within the amplifier.

Both of the above types of non-linear distortion can be eliminated by reducing the size of the signal, but it is possible for an amplifier to have non-linear distortion even when there is no overloading. This is because no amplifying device is perfectly linear, the gain always changes when the signal amplitude is changed. Non-linear distortion changes the shape of a sinewave, but has no effect on the shape of a squarewave. This is the opposite of the effect of frequency distortion.

Feedback

Imagine an amplifier *A*, and a network *B*, which might be resistive, integrating, differentiating etc., connected as in Fig. 2.5. The effect which the amplifier and the network have on each other depends very much on whether the amplifier is inverting or non-inverting.

When a network connects the output of an amplifier back to its input, it is called a *feedback network*, and the circuit from output back to input is called a *feedback loop*. If the amplifier is inverting,

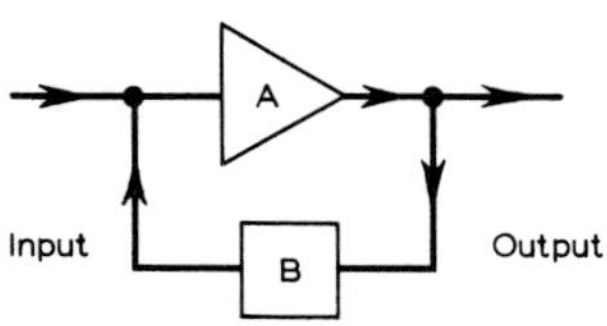

Fig. 2.5 Principle of feedback.

meaning that the output from which the feedback is taken is inverted relative to the input to which the feedback is brought, the feedback is said to be negative. If the amplifier is non-inverting, the feedback is said to be positive.

Positive Feedback

Suppose, in the network of Fig. 2.5, the amplifier has a voltage gain, allowing for input and output impedances, of +20 and the feedback network has a loss of 10. A signal of 1V at the input would be amplified to 20V at the output and reduced to 1/10 of 20V, which is 2V, by the feedback network. This feedback signal is greater than the applied signal and is in the same phase and being added to the input. The amplifier can now supply its own input signal. A system such as this *oscillates*, the output amplitude rises as fast as time constants allow, overloads, falls again as fast as time constants allow, overloads in the opposite direction and so on.

If we make the feedback network of such a system in the form of a time constant, the time between rising amplitude and falling amplitude will be controlled by the time constant, and the oscillator is called a relaxation oscillator, since after a rise in amplitude the oscillator does nothing until the time constant allows the amplitude to fall again. If the feedback network is a tuned circuit, then the output of the oscillator may be a sinewave of the frequency to which the feedback circuit is tuned. We shall see later that the waveform is a sinewave only if other requirements are also fulfilled.

Oscillation occurs only if the amplifier can supply its own input, i.e. if the loss of the network is not quite so much as the gain of the amplifier. If the loss of the feedback network is greater than the amplifier gain, then the positive feedback has the following effects:

(1) Since the feedback signal adds to the input signal, gain is increased.

(2) Because the signal fed back has already been distorted by the amplifier, distortion is increased.

(3) Because noise generated in the amplifier is fed back, noise is increased.

(4) Because the G-B factor is constant and gain has been increased, bandwidth is less.

Positive feedback is used:

(a) when very high gain is required with few amplification stages,

(b) when the rise times of an amplifier have to be improved regardless of other factors (by feeding back through a differentiating network),

(c) when an amplifier must pick out and amplify one frequency more than any other (feedback network is a tuned circuit passing that frequency),

(d) when an amplifier must reject a frequency in favour of others (feedback network is a tuned circuit rejecting that frequency),

(e) when the input impedance of an amplifier is to be increased. When used in this way, the use of positive feedback is often called 'bootstrapping'.

An example of bootstrapping in action is shown in Fig. 2.6. The output voltage is used to provide a feedback in phase and almost equal to the input voltage at the other end of the input resistor. This causes both ends of the input resistor to change voltage in phase and with practically equal amplitude, so that practically no current flows in the input resistor.

This is equivalent to making the input resistor of a very high resistance value for signal voltages, while behaving normally for d.c. The current in the input resistor is given by:

$$\frac{\text{input voltage} - \text{feedback}}{\text{actual value of resistor}}$$

and so the resistor appears to have the value

$$\frac{\text{signal voltage}}{\text{current flowing}}$$

which can be very large compared with its actual value for d.c.

The 'gain' in resistance can also be found by taking the feedback 'loop gain' (the gain from input to the other end of the feedback loop). If this is X, then the resistance is multiplied by $1/(1-X)$. For example, if the gain of an amplifier from its input to the other end of the input resistance is 0·9, then the input resistance value appears multiplied by $1/(1-0{\cdot}9)$, which is 10.

Negative Feedback

When feedback is applied to an inverting amplifier, the signal fed back is subtracted from the input signal. This has several important effects:

(1) since the fed back signal is subtracted, the gain of the whole system is less,

(2) because amplifier distortion is subtracted, distortion is less,

(3) because noise generated is fed back, noise is less,

(4) because G-B factor is constant but gain is less, bandwidth is increased.

To see what scale of change is caused, imagine an amplifier of amplification A. For one volt fed in at the input, the voltage of signal at the output will be A volts. If a fraction, X, of this output is fed back, then the voltage at the input becomes less. Imagine the feedback network connected, so that one volt is fed in and a voltage V appears at the output. V will now represent the gain of the amplifier with feedback. The amount fed back will be $X \times V$, so

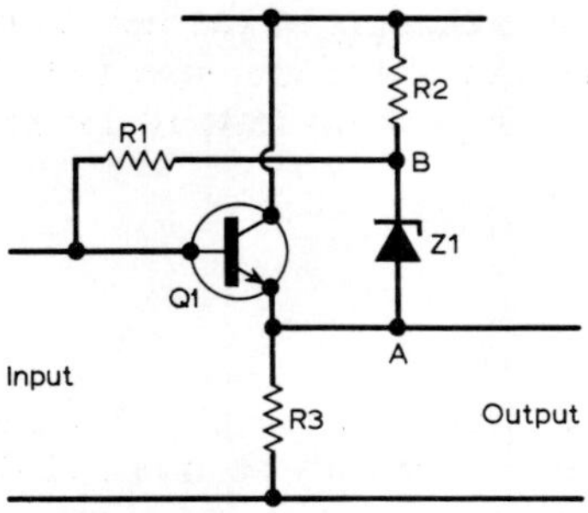

Fig. 2.6 Bootstrapping. The signal voltage at point B is almost exactly equal to that at A, but a few volts higher in d.c. level due to the Zener diode Z1. This ensures a steady bias voltage across R1, but practically no signal voltage.

that the actual input to the amplifying part will be $1-X\times V$, the one volt fed in less the amount fed back.

Now the gain of the amplifier A acts on the voltage $(1-X\times V)$ to produce the output V, so that $V=A(1-X\times V)$. Solving this gives $V=A/(1+A\times X)$, which is obviously less than A. Note that some texts take X as being negative, so giving a negative sign between 1 and $X\times V$. The fraction $1/(1+A\times X)$ expresses by how much gain is reduced and also by how much distortion, noise etc. are reduced.

Effect of Feedback on Response

Important as these effects may be, there is another point of even greater importance. When a feedback network is used with an amplifier of high gain, the response of the whole system becomes almost exactly the response of the inverse of the feedback network. Some examples may make this clear.

Suppose that we have an amplifier with a gain of 120dB. To it we attach a feedback network of 10dB loss, a resistive potential divider. The gain of the amplifier is now 10dB with the network attached. If the feedback network had a 5dB loss, the gain of the amplifier plus network would be 5dB. If the feedback network were an integrator, the response of the amplifier plus network would be that of a differentiator; and if the feedback network differentiated, the whole system would integrate.

All this holds good providing that the gain of the amplifier is high compared to the loss of the feedback network (at least ten times more) and providing that the feedback stays negative despite phase shifts due to time constants. In this way, the response of active devices can be controlled exactly with the passive components in the feedback loop and the gain and characteristics of the complete amplifier remain constant despite changes in the behaviour in the active devices so long as the gain of the amplifier is high compared to the loss of the feedback network (at least 10:1); and that the feedback ratio is constant.

Feedback Troubles

The application of negative feedback has been one of the largest steps ever made in circuit design and has certainly been one of the most abused. Many designers have treated it as a universal remedy for all amplifier faults, which it is not, and as a method for making poorly designed amplifiers behave, which it does not. In particular, the thoughtless application of negative feedback can cause trouble when:

(1) The amplifier or the feedback loop contains more than two time constants.

(2) A transformer is included in the amplifier or feedback loop.

(3) A tuned circuit is present in amplifier or feedback loop.

In such cases, a well designed amplifier can be completely stable, i.e. free from oscillation due to the negative feedback being shifted positive by phase-shifting circuits; a poorly designed amplifier may oscillate continuously at high frequencies. In all cases where feedback amplifiers give trouble which seems due to oscillation, varying gain, ready overloading, or high noise, the service engineer should suspect instability and check by turning down the input to the amplifier and disconnecting the feedback network.

If the trouble ceases (remember that the gain of the amplifier is now higher) then phase shifting has been the trouble. If the phase shift of the amplifier can be measured at the most troublesome frequency (that of highest gain or oscillation) the feedback network or the amplifier time constants must be adjusted so that the phase shift is reduced.

Stability

An amplifier which is stable with a resistive load may be unstable with an inductive or capacitive load. If an amplifier which appears perfect on test gives trouble in service the load should be suspected. Feedback amplifiers are particularly liable to be troublesome in this way. As advised above, when a feedback amplifier gives trouble, check the amplifier with the feedback loop disconnected. In many cases the gain of an amplifier may fall drastically with little effect on the whole system, since the feedback system determines the overall gain. Eventually the amplifier gain becomes so low that the feedback system can no longer compensate.

Matching

When we discussed input and output impedances of amplifiers, we said that for a voltage amplifier it was desirable to feed the signal from a source of low impedance, so that the potential divider formed by source impedance and amplifier input impedance had least effect on the signal. For the same reason, the voltage amplifier should feed into a high impedance. Source and load impedances are not always of the desired value, and any measures taken to make them so are termed *matching*.

Matching is particularly important in the case of power amplifiers where the power gain is more important than the voltage gain. In this case, power gain is greatest when the source impedance equals the amplifier input impedance and the load impedance equals the output impedance. Matching is also important when pulse amplifiers are connected by long cables, or when output stages of transmitters are connected to aerials.

In both cases, the cables have a *characteristic impedance* (TV coaxial cable has an impedance of 75Ω) which has to be matched to amplifier, aerial, or any other device connected to the cable. Matching may be done by adding series resistors (to increase an impedance) or by shunt resistors (to decrease an impedance) or by the use of transformers. Where transformers are used, the ratio of impedances should be the square of the ratio of turns; for example, 10kΩ matched to 100Ω is the ratio 10,000/100 = 100, the turns ratio must be 10, since $10^2 = 100$.

Both input and output impedances are altered by feedback, depending on the type of feedback and the way in which it is connected. Positive feedback, as normally used, raises impedances; negative feedback as normally used reduces output impedances. Negative feedback connected directly to the input of an amplifier *reduces* input impedance; indirectly connected (for example, if input is on a transistor base and feedback is connected to the emitter) it *raises* input impedance.

Noise

In all amplifiers, the voltage present at each stage consists of the wanted signal plus unwanted random changes of voltage called *noise*. Noise voltages or currents can be reduced by careful selection of components, but there is a minimum value which can be reduced only by lowering the temperature of the components of the amplifier.

The noise at the input of an amplifier is the most important, since it is amplified as much as is the signal. The total noise voltage depends on the bandwidth as well as on the temperature, and can therefore be made less troublesome if the bandwidth of the amplifier can be limited. Noise contributed by an amplifier may be expressed either as a noise figure or as an 'equivalent noise resistor'.

The noise figure is the ratio of noise, in r.m.s. volts at the input of the amplifier (for a quoted bandwidth), to the noise created by the resistance of the signal source (Johnson noise) for the same bandwidth. The Johnson noise voltage is given by $\sqrt{4kToRo\Delta f}$ where $k = 1{\cdot}38 \times 10^{-23}$, To is the temperature in Kelvins (°C+273) and Ro is the source resistance and Δf is the bandwidth between the 3dB points. Noise figure may also be given in dB as 10 log $\sqrt{4kToRo\Delta f}$ or 5 log $(4kToRo\Delta f)$.

When a noise resistance is quoted, the value of resistance is quoted which would generate Johnson noise of the same value as the amplifier noise. In some cases, a noise spectrum is shown; a graph of noise against frequency, plotted in the form: noise voltage/ $\sqrt{\text{bandwidth}}$, again because this gives a simpler shape of graph and conveys more useful information.

Operational Amplifiers

The word 'operational' was originally used to describe an amplifier of very high gain which could be modified by feedback to carry out various mathematical processes in analogue computers. The word is now used for any high-gain direct-coupled amplifier and, in practice, mostly for integrated amplifiers. Because operational

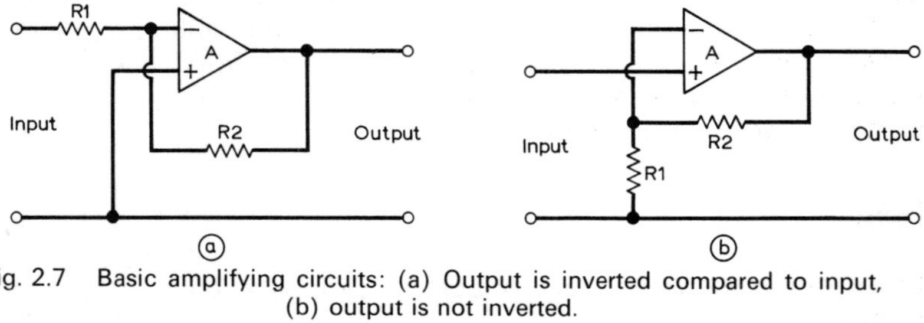

Fig. 2.7 Basic amplifying circuits: (a) Output is inverted compared to input, (b) output is not inverted.

amplifiers are now available in integrated form for the same price as a couple of transistors there is considerable incentive to design equipment round these readymade components.

Most operational amplifiers (known usually by the abbreviation op-amp) have one output and two input terminals in addition to 'earth'. One input provides inversion and gain, the other the same gain without inversion, and it is the difference in signal voltage between the two input terminals which is amplified. Ideally, an op-amp should have very high gain (so that its performance can be regulated by the feedback network connected to it) and very high input impedance so that the amplifier does not disturb the working of the stage which feeds it.

It should have a large bandwidth from d.c. up to very high frequencies with only very small phase changes, low output impedance so that its performance is unaffected by changes in the load connected to it, and minimum possible offset, so that the output signal will be zero when the input (d.c.) signal is zero.

The two basic amplifying circuit connections are shown in Fig. 2.7. If the open-loop gain of the op-amp (the gain without feedback) is very large, and the resistor *R*2 is not too large (perhaps not more than 100 times) compared to *R*1, then the gain of the amplifier is given to a very close approximation by *R*2/*R*1. The exact expressions are shown, but the approximation is close enough for most purposes.

CHAPTER THREE

LOW FREQUENCY AMPLIFIERS

IN ELECTRONICS, amplifiers may be required for any frequency of signal which is available, but by far the greatest number of untuned amplifiers are built for the frequency range between 10Hz and 20kHz and are termed 'low frequency amplifiers'. These may be subdivided into voltage amplifiers, current amplifiers and power amplifiers according to which feature is most desired at the output (though it is possible that all three might be wanted).

For example, if the amplifier forms part of an oscilloscope display where a large signal voltage must be applied to the deflector plates of a cathode ray tube, then a voltage amplifier is wanted, because the current and power of the output is irrelevant to the need of the circuit. Similarly, if an amplifier is required to make a 10mA meter read full scale for 100μA of a.c., a current amplifier is required. If the amplifier is required to operate a device such as a loudspeaker, a vibrator or a solenoid, then a power amplifier is required.

L.F. Voltage Amplifiers

The simplest voltage amplifier is a transformer, but no amplification of power is possible. Although it is possible to construct transformers to deal with the range 10Hz to 20kHz, large step-up ratios are not easily attained with such a frequency range (because of stray capacitance and leakage inductance) if any significant amount of power has to be transferred as well. For this reason, and also because of the size of a transformer (although very small transformers are available), voltage amplification is seldom now carried out using transformers alone, and their use is diminishing even in conjunction with other circuit elements.

Fig. 3.1 shows simple amplifier circuits using valves and transistors. Fig. 3.1(a) shows a one-transistor amplifier. With no signal applied, a current of approximately 1mA passes through the transistor, and the bias components *R*1, *R*2 and *R*3 ensure that

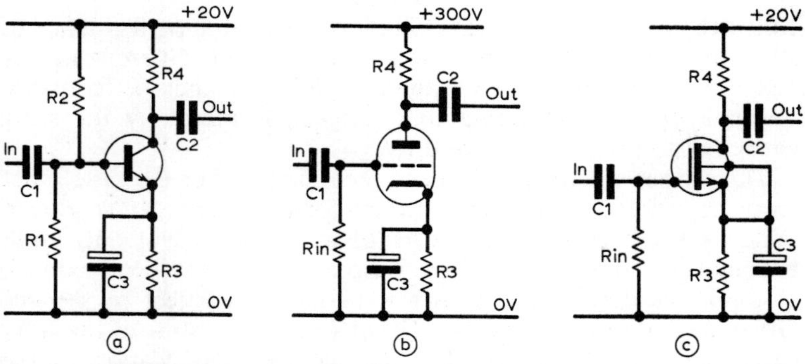

Fig. 3.1 Simple single-stage amplifiers: (a) n–p–n transistor, (b) triode valve, (c) p-channel MOSFET. In the transistor amplifier, Rin consists of R1 and R2 in parallel with the input resistance of the transistor.

this bias current is maintained despite temperature changes. *C*1 is the capacitor used to couple the input signal into the amplifier, and *C*2 is used to couple the signal out from the collector. *R*4 is the collector load resistor; changes of current in the collector circuit cause changes of voltage across this resistor and these changes of voltage constitute the output signal.

In the circuit as shown, the performance is limited by the time constants used and by the behaviour of the active device, whether valve, transistor or FET. When the capacitor *C*3 is omitted, the gain becomes less because some of the signal current in the active device is being used to develop a voltage across *R*3, and the input to the amplifier is less by this amount (assuming voltage drive from a low source resistance). The proviso in brackets is needed because when the stage is driven from a high source resistance the voltage across *R*3 will cause very little change to the input current.

At the same time, the input resistance of the amplifier is greater by the inverse of the feedback factor (meaning that if the gain is halved, the input impedance is doubled) and the output impedance of the active device is increased. It should be remembered, however, that the output impedance of the active device is in parallel with the load resistor as far as the next stage is concerned, so that the change of impedance is less noticeable. In addition, the distortion caused by the stage decreases and the frequency range is increased. These effects are due to the negative feedback caused by the voltage developed across *R*3 being subtracted from the input voltage.

The gain of the simple amplifier is usually calculated and measured at a signal frequency of 1kHz (by convention) and all changes of gain with frequency are referred to the gain at this frequency, called the mid-frequency gain. The low-frequency performance of this amplifier is set by the time constants which differentiate the signal ($C1 \times R_{in}$ and $C3 \times R3$) and the high frequency performance depends on the integrating time constants caused by the stray capacitances across $R1$, $R2$ and $R4$.

The output stray capacitance is generally smaller than the input stray capacitance, which has to be corrected for *Miller Effect*. When a small capacitance exists between the output and input terminals of an amplifier whose output is inverted, then feedback through the capacitance causes it to have the effect of a much larger capacitance of value $A \times C$, where C is the value of the stray capacitance by itself, and A is the voltage amplification of the stage.

In transistor amplifiers, this inflated value of input capacitance is seldom troublesome because of the low values of input resistors used in most circuits. In valve or FET amplifiers it may cause trouble which has to be overcome either by using a device with internal screening (pentode) or by the use of circuits which cancel the effect (bootstrapping). The Miller Effect is reduced if series feedback is used by omitting the emitter or cathode bypass capacitor. There are two main sources of stray capacitance:

(1) the capacitances which exist across the terminals of a transistor or valve and which are quoted in manufacturers' handbooks. These will be 'typical' average values, and will, especially for the transistor, vary with working conditions. Allowance will have to be made for Miller Effect.

(2) The stray capacitances which exist in printed circuit or other circuit boards, valve retainers, and wiring. These will depend on the positioning of circuit wires or lines and can be the cause of two identical circuit diagrams giving rise to very different amplifier performance.

Impedance and Gain

The input impedance of the simple l.f. amplifier at mid-frequency depends on the device used. In the case of a valve, the input impedance is practically that of the grid resistor, provided that the applied signal does not drive the grid more positive than the cathode. Junction FETs have similar orders of input impedance; MOSFETs have very high impedance for signals of either polarity relative to the source. For a transistor in the simple circuit of Fig. 3.1, the input impedance is typically 5kΩ for a modern silicon type, less than 1kΩ for an older germanium type.

The impedance can be raised to some extent by using very low bias currents; for example the input impedance of a general-purpose silicon transistor can be more than 100kΩ when the collector current is less than 0·1mA. A further increase can be achieved by omitting the emitter bypass capacitor, but in general the input impedances are not as high as those easily obtainable with simple valve or FET circuits.

The output impedances of the amplifiers are set approximately by the series resistance in the equivalent circuit in parallel with the load resistor. The resistances of semiconductors and of triode valves are within the range 10kΩ to 100kΩ. Pentode valves have very high output resistances, so that they may be thought of as constant current devices, because a small change of voltage at one end of a large resistance has little effect on the current flowing through it.

The gain of the simple amplifier is fairly easily calculated if it is assumed that the amplifier is fed from a signal source of almost zero impedance and is feeding into a load of very high impedance. This is practically never true, and when a two-stage amplifier is made by coupling two simple amplifier stages together (a process called cascading), the gain is not as much as might be expected, especially in the case of transistor amplifiers, which are more affected by input and output impedances than are valve amplifiers.

In a cascaded amplifier, the load into which one stage works is the input impedance of the second. At mid-frequency, the impedance of all capacitors is negligible, and the supply line is at the same *signal* voltage as earth, since it is coupled to earth at the power supply by a large capacitor. It is for this reason that the load resistance can be imagined as being in parallel with the output impedance of the transistor.

Calculations of gain for a cascaded amplifier are seldom very accurate, and rough methods are good enough for most purposes. Where an amplifier of exact gain is required, designers use negative feedback to set the gain of an amplifier whose open-loop gain is very high, so that the gain of the whole amplifier is the inverse of the loss of the feedback loop.

Common Electrodes

The amplifiers of Fig. 3.1 are termed common-cathode, common-emitter or common-source amplifiers because the input signals are fed in between an electrode and the common electrode, and the output signal taken between an electrode and the common electrode.

In the circuits shown, the cathode/emitter/source is common to both input and output circuits. Though this is the best circuit from the point of view of high gain in a simple amplifier, it is not the only possible arrangement.

Confining remarks to transistor circuits only, the common-collector circuit of Fig. 3.2 is much used for its high input impedance (1MΩ or more) and low output impedance (a few ohms), though the voltage gain is always less than 1. The transistor is still amplifying, but the input signal to the active device is the difference between the output and the input signals if the transistor is being driven from a comparatively low impedance. The greater the gain of the transistor, the smaller the input needs to be for a given output, and so the closer the gain of the whole circuit approaches a value of 1.

This circuit is known as the *emitter follower* (cathode follower or source follower for the other devices) because the emitter of the transistor has a signal output which follows the input at the base, providing that time constants permit it. The output is not inverted. The circuit may be used for matching from a high impedance to a low impedance (note that it offers *current* amplification) and for isolating stages.

The common base circuit of Fig. 3.3 is seldom used in l.f. amplifiers, but is useful at high frequencies. The input impedance is low, a few ohms and the output impedance high, several hundred kΩ, and the gain roughly half as much as that of the common emitter circuit. Note that the output is not inverted.

Failure of L.F. Voltage Amplifiers

L.F. voltage amplifiers may fail in a variety of ways:

(1) The gain of the amplifier may be reduced to an unacceptable level,

(2) the frequency response of the amplifier may become too narrow,

(3) the output signal may become distorted,

(4) the amplifier may oscillate,

(5) a large amount of the output signal may consist of noise,

(6) there may be no output at all.

The last case is most easily dealt with, as it implies a total breakdown of a component or components, but the other symptoms can be more baffling.

Gain: When the gain of an amplifier is reduced this may be due

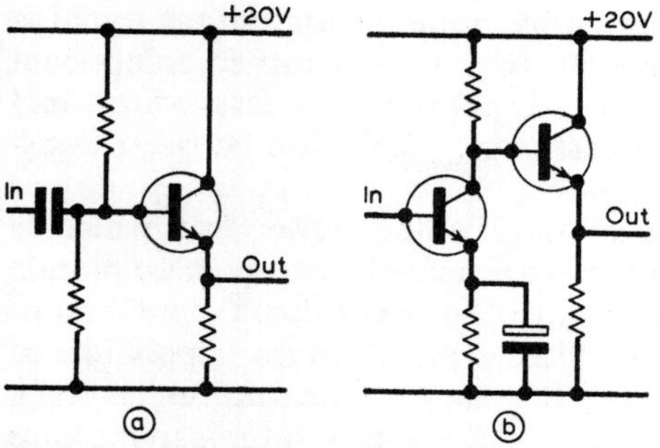

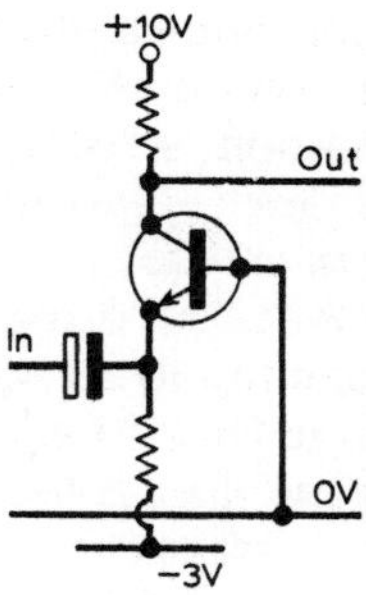

Fig. 3.2 (left) Common-collector circuit, with two methods of biasing. (b) shows the common-collector circuit coupled directly to a common-emitter amplifier stage.

Fig. 3.3 (right) Common-base stage. An alternative bias system consists of biasing as for a common-emitter stage, but with the base decoupled to earth and the signal input to the emitter.

to failing active devices, especially if valves are used, load resistors reducing in value (common if the load resistors are running at high dissipation) or emitter bypass capacitors becoming open-circuit. In addition there is the possibility in a feedback amplifier of a change in the components in the feedback network. In some circuits a change in bias can cause a considerable change in gain, but the usual effect of bias change is increased distortion rather than reduced gain. In valve circuits, the supply to, and the decoupling of, screen grids can cause considerable gain changes.

Distortion: Distortion in an amplifier may be caused by excessive input signal, reduction in the bias of one or more amplifier stages, saturation in a coupling transformer, or disconnection of a resistor carrying d.c. to a stage. In an emitter follower, loss of supply to the collector has little effect on signal amplitude but can make the output much less linear. Even the disconnection of a load resistor in a feedback amplifier may have less effect on gain than on distortion.

Frequency Response: When the frequency response of an amplifier deteriorates, this is due to a change in the time constants. At the low frequency end of the range, coupling components may have changed in value, as also may emitter time constants. At the high frequency end stray capacitances may have increased due to disturbance in the layout of components (sometimes, alas, due to previous servicing) or load resistors increased in value, the most likely cause which also shows up as increased gain.

Oscillation: This is usually a disease of feedback amplifiers, caused by a change in the time constants, but it can occur in high-gain, non-feedback, amplifiers when some of the output signal is picked up at the input. This may be due to disturbed component location, screening plates removed, poorly fitted earth wires, bad soldered joints (especially in earthing), interaction between transformers etc.

Another cause of feedback may be the power line which is common to all stages and can carry a signal from a high amplitude stage back to a low amplitude stage. To avoid this, the method of decoupling is used; a low pass filter is placed in the supply line to early stages or in the supply to each stage. Some caution has to be used here, for the effect of a simple *CR* low pass filter is to increase the gain of the amplifier at very low frequencies, so making it easier to oscillate at those frequencies. A more certain cure is to use a stabilised power supply. When oscillation occurs in an amplifier previously known to be stable, faulty decoupling or power supply stabilisation (or even smoothing capacitors) should be suspected.

In high gain amplifiers, badly placed earthing points may also cause oscillation due to currents flowing in the earth return path which is common to both early and later stages. Output stages should have their earth returns taken directly to the earth end of the smoothing capacitor in the power supply, and earlier stages should also be linked directly to this point rather than linked through any other stages.

Feedback amplifiers which oscillate should have the feedback loop disconnected. If the oscillation ceases, then the feedback loop is faulty or the phase shift in the amplifier is excessive. Sometimes the design may be so unstable that oscillation at one frequency takes place, because of the high gain, with the feedback loop removed; with the feedback loop connected the oscillation is at another frequency. In such a case, the phase response of the amplifier has to be plotted a stage at a time and modified so as to prevent the trouble.

Noise: Noise in l.f. amplifiers is usually of two main types—broad band noise sounding in a loudspeaker as a rushing sound and appearing on the oscilloscope as a jagged irregular trace, and mains hum either at mains frequency or twice mains frequency (as whole wave rectification doubles the frequency). When a feedback amplifier develops noise suddenly, this is often due to disconnection of the feedback loop; otherwise, broadband noise in practically every case originates in the input stage.

The first transistor in the amplifier may be noisy, possibly due to incorrect bias causing too high a collector current. The resistors in the input stage may be noisy—this can be checked by connecting a

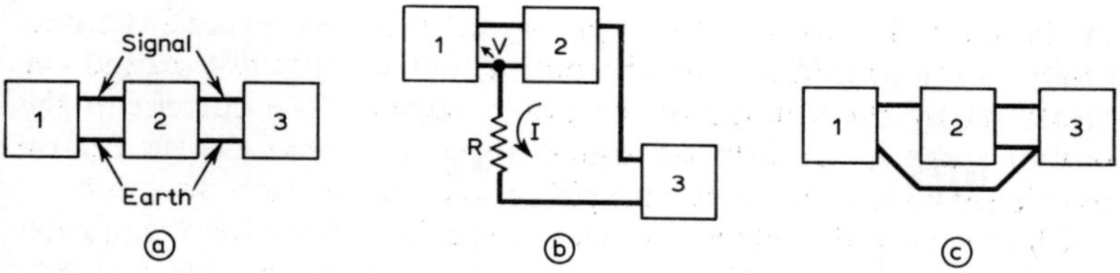

Fig. 3.4 Earth loops: (a) low-voltage stage 1 coupled to a high current (output) stage 2 and a power supply 3; (b) the loop redrawn. The current from 2 to 3 can now be seen to cause a signal voltage V1 to stage 1; (c) The solution: separate earth leads avoid the common impedance.

high value capacitance across each in turn. In high gain amplifiers, only low noise resistors should be used in input stages.

Mains hum is often very different to cure. In the simplest case, failure of smoothing components may be causing ripple in the supply. The input of the amplifier may be o/c, so picking up the 50Hz mains frequency which is radiated anywhere near mains supplies. The most common cause, however, is the earth loop, mentioned earlier under the heading of 'Oscillation'. Fig. 3.4 shows the effect of having a portion of an earth wire common to several stages. The hum which would be unnoticeable in the later stages is picked up and amplified by the early stages because of the common coupling. The remedy is also shown.

L.F. Power Amplifiers

A power-consuming load, such as a loudspeaker, heating coil, aerial or relay requires power to be delivered to it from an amplifier. Such loads have an impedance, and to deliver maximum power, the amplifier should have an output impedance of the same value (but see note on transistor power amplifiers later). In some cases this requires matching by means of a transformer, in other cases direct coupling can be used. An l.f. power amplifier inevitably consumes power, ranging from an amount little more than the power being delivered up to about five times this amount depending on the type of circuit used.

Class A Power Amplifiers

A Class A l.f. power amplifier takes a steady current (averaged over a cycle) from the power supply and delivers a varying power to the load. To do this, the input signal must not exceed the bias limits of the amplifier, and the output signal must not exceed the limits set by the voltage of the power supply. The current in the output stage is set with no signal applied so that the maximum usable signal just drives the amplifier to cutoff in each direction.

Class A amplifier stages are used when low distortion is required and the load is variable, and it is an advantage to have a steady load on the power supply. Their efficiency, that is:

$$\frac{\text{Power delivered to load}}{\text{Power taken from power supply}}$$

can never be more than 0·5 (50%) and is usually considerably less. Two Class A amplifiers are shown in Fig. 3.5. The valve amplifier uses transformer coupling to feed a vibrator; the transistor amplifier feeds a loudspeaker through a capacitor. Maximum power output to the load of the valve amplifier is attained when the load impedance is equal to the amplifier impedance; for this reason the valve with an output impedance of 20kΩ is coupled to the vibrator, impedance 20Ω, by a transformer whose ratio is $\sqrt{20{,}000/20}$ which is approximately 30:1.

The transistor amplifier is also matched to the loudspeaker, in this case by using a loudspeaker of fairly high impedance. With both circuits shown, a loudspeaker impedance which is decidedly more or less than the matching impedance will draw less power from the amplifier. This is not the case for direct coupled push-pull Class B transistor amplifiers (see later), most of which deliver power which depends on the supply voltage as well as on the load resistance. The lower the load resistance, the greater the power which the amplifier can deliver within the limits of current which the transistors can handle.

The biasing of Class A amplifiers must be arranged so that current never cuts off nor becomes so high that the quantity (load impedance × peak current) could exceed supply voltage (except in transformer coupling). When transformer coupling is used, the bias current causes practically no voltage drop across the primary, the maximum signal current flowing causes the voltage to drop to nearly earth potential at the collector/anode; at minimum signal current the voltage rises to twice the supply voltage.

This happens because of the bias current leaving the voltage at about supply voltage—so as to preserve this as the average voltage, the volts must rise to 2 × supply at minimum current. When transformer coupling is not used, or when the load is totally non-inductive, the bias current is chosen so that half of the supply voltage is dropped across the load resistor.

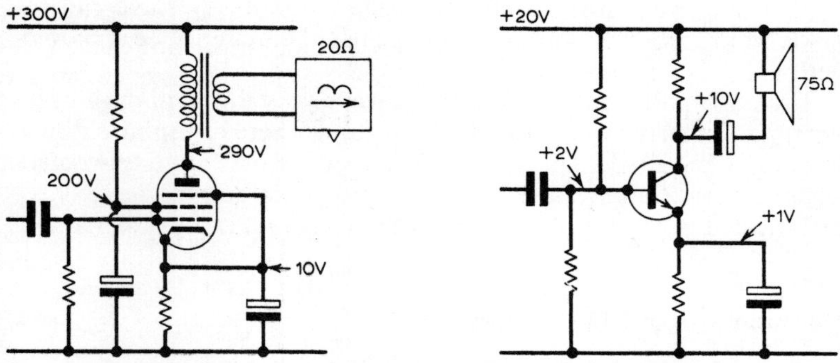

Fig. 3.5 Simple l.f. power amplifier, showing typical voltages.

Push-pull Class A amplifiers use two valves or transistors feeding a common output. The active components are connected so that a positive-going signal into one device gives the same output as a negative-going signal into the other. Fig. 3.6 shows typical circuits of this type. The name is derived from the idea that when one device has increasing current the other has decreasing current; one is pushing, the other pulling, at any given time. The term 'balanced circuit' is also used.

Advantages of the Push-Pull Circuit

(1) If the same input signal is fed to both inputs, there is no output. This is because one input must be inverted to produce an output in the same phase. This feature makes the amplifier insensitive to hum or noise picked up at both inputs in phase.

(2) The output is more linear. If the two devices are matched so that they distort the signal in the same way, at least some of the distortion is cancelled out.

(3) In the circuit of Fig. 3.6(a), the d.c. in the two halves of the transformer primary causes the magnetic fields due to d.c. to cancel, and the core of the transformer is not continually magnetised. A smaller transformer can therefore be used without danger of saturation than would have been the case with a single output device passing the same current.

(4) In the circuit of Fig. 3.6(b) the output impedance is low, and can be low enough with transistors to feed a loudspeaker without using a matching transformer. Pulses of either polarity can be obtained at the output without the time constant distortion which occurs with single transistor stages.

(5) Using transistors, a complementary symmetry stage can be made which requires only one input signal. This cannot be achieved with valves. (Fig. 3.6(c)).

The push-pull principle is also extensively used in voltage amplifiers, particularly in integrated circuits and where a response down to d.c. is required. The main advantage here is that unwanted signals in the same phase at each input (common-mode signals) are rejected in favour of the differential (push-pull) input signal.

Disadvantages of Class A Operation

(1) There is a steady bias current in the active device or devices. This is tolerable in valves, less so in silicon transistors, hardly tolerable in germanium transistors.

(2) The efficiency is low. Other circuits using transistors can achieve very much higher efficiency. Because of the lower efficiency, a Class A transistor amplifier requires much greater heat sink area than the more common Class B design, and the high standing current from the power supply makes smoothing more difficult.

In high quality work, the differences become less important. Stabilised power supplies are used, ensuring that a Class A design can be supplied with a well smoothed supply, or a Class B design with a supply which can accommodate fluctuating currents. In either case large amounts of feedback can be used to make distortion figures very low. Distortion figures of below 0·05% are now common in the high quality amplifiers working at audio frequencies and delivering one hundred watts or more.

(3) A current is taken from the power supply whether or not any signal is being amplified. This may be of importance in battery operated equipment.

Class B Amplifiers

In a Class B amplifier, the bias of the active device is deliberately reduced. This would lead to severe distortion in a stage using a single active device, and so the Class B stage is always found as a push-pull stage. The idea behind this is that the two active devices behave as a push-pull pair when both are conducting, and when one cuts off the other is conducting. Fig. 3.7 shows the principle.

For small signals, the current drawn from the power supply is steady, but when the input signal is enough to cut off each active device alternately, the total current drawn from the supply in-

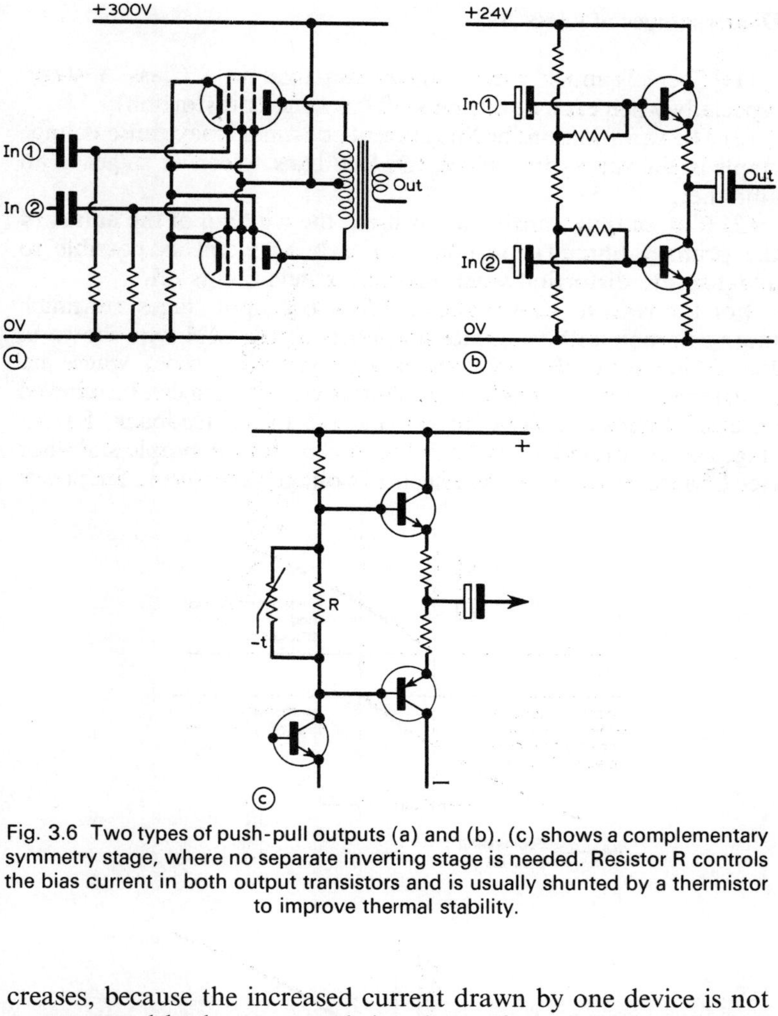

Fig. 3.6 Two types of push-pull outputs (a) and (b). (c) shows a complementary symmetry stage, where no separate inverting stage is needed. Resistor R controls the bias current in both output transistors and is usually shunted by a thermistor to improve thermal stability.

creases, because the increased current drawn by one device is not compensated by less current being drawn from the other if it has cut off. The advantages of Class B amplifiers are as follows:

(1) Small bias current drawn from the power supply.

(2) Current drawn is proportional to the signal power.

(3) Efficiency is high (70–80%).

(4) Low dissipation in active devices when there is no signal input; dissipation increases as signal increases.

These advantages have made the use of Class B very common in transistor amplifiers.

Disadvantages of Class B

(1) There is much greater distortion than in a Class A stage, especially when each device cuts off (cross-over distortion).

(2) The variations in the line current consumed may cause voltage ripple in the supply line which can feed back to earlier stages in an amplifier.

(3) If an output transformer is used, the winding of the halves of the primary (circuit) must be as nearly identical as possible to avoid severe distortion when one active device cuts off.

For the reasons shown above, Class B output stages are much more common with transistor amplifiers of the LIN type shown in Fig. 3.6(c) (and the complementary symmetry type) where no transformer is used. Much of the distortion problem can be relieved by careful choice of bias current and the use of feedback. Earlier stages of an amplifier can be fed with d.c. from a simple stabiliser (see Chapter 9) to avoid the effects of voltage variation in the power

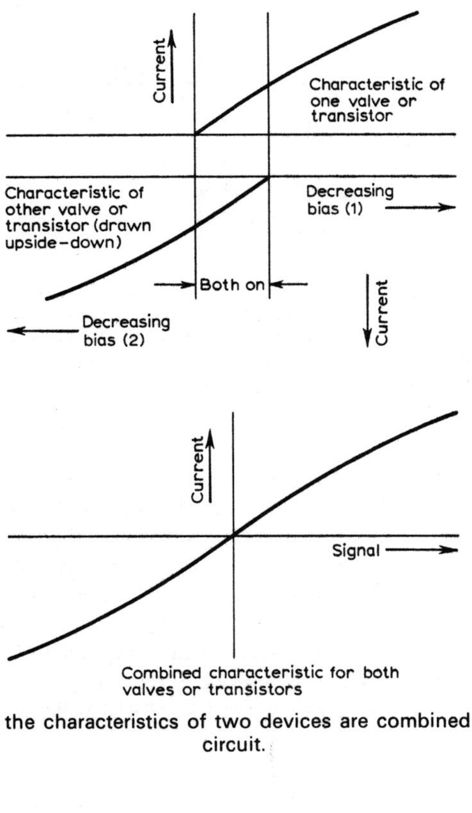

Fig. 3.7 How the characteristics of two devices are combined in a push-pull circuit.

supply. In addition, the Class B stage is a push-pull stage and has all the advantages of push-pull operation.

Failure of Power Amplifiers

(1) **No signal out:** Signal(s) in or out disconnected.
No power supply.
Active device(s) biased off.
Active device(s) has/have failed (unlikely).

(2) **Power output very low:** Input signal low.
Bias current too high or too low.
Load wrong impedance.
Power supply inadequate.
Low emission valves (if used).
One device of push-pull pair has failed.
S/c turns in output transformer.

(3) **Severe distortion:** Insufficient current or voltage from power supply.
Excessive input signal.
Bias incorrect.
One device p/p pair has failed.
(push/pull) s/c turns on one half of transformer primary.

(4) **Severe overheating of transistors:**
Class A—bias current too high
output transformer faulty.
Class B—bias current too high
output s/c or low impedance.

Paraphase Outputs or Inverted Outputs

When a push-pull output stage is used, it is necessary at some stage to split the signals in the amplifier into two portions, identical except that one is inverted compared to the other. An amplifier which accomplishes this is called a *paraphase* or an *inverting* amplifier. This technique finds other uses. For example, if a signal is obtained in the form of two components, one inverted, it can be amplified despite interference from hum and pick-up because these unwanted signals are not generated or picked up in paraphase. Such an amplifier is termed a *balanced* amplifier and is said to *discriminate* against in-phase (unwanted) signals.

Paraphase circuits are shown in Figs. 3.8. The first uses equal loads at both the anode and cathode of a valve, or at the collector and emitter of a transistor. Since, with a triode or a transistor, the

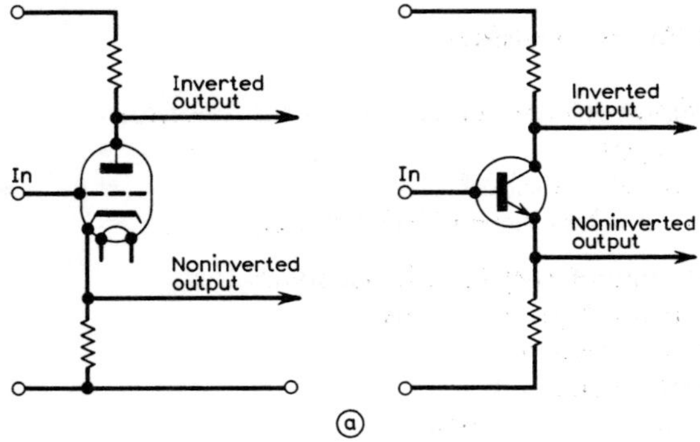

Fig. 3.8 Paraphase circuits: (a) simple paraphase, (b) methods of biasing.

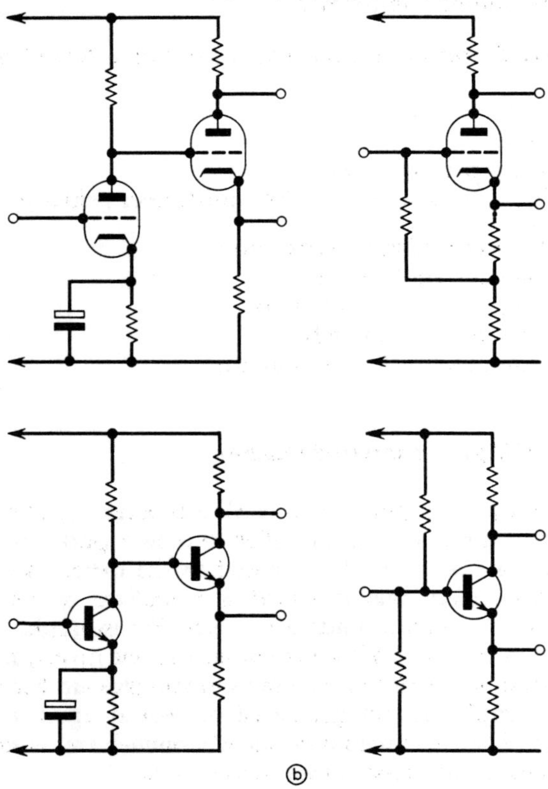

same signal current flows in the anode/collector as in the cathode/emitter (if we disregard the current taken by the base) the two signals out must be equal in size. They are opposite, however, because increasing current causes the cathode/emitter voltage to *rise* and the anode/collector voltage to fall.

The slight loss of balance in a paraphase of this type using a transistor can be compensated by making the emitter resistor slightly *smaller* (as it takes the base current as well) than the collector resistor. Note that biasing this circuit requires some care to ensure sufficient bias current through the active device and that the output impedances are unequal and that the gain to each load is *unity*.

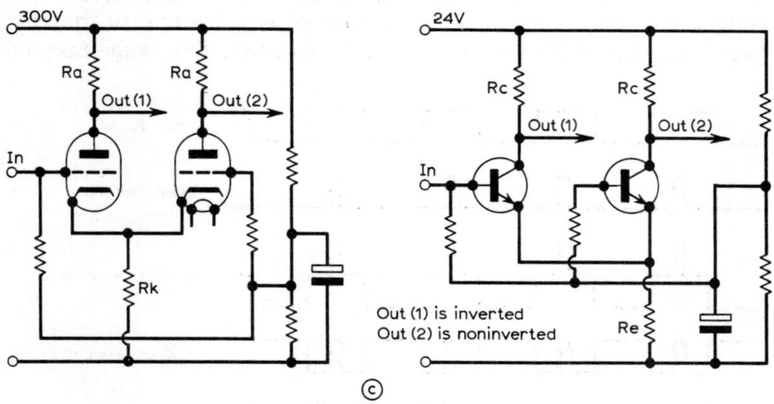

Fig. 3.8 (c) Paraphase circuits: long-tailed pair.

Fig. 3.8(c) shows the transistor and valve versions of the *long tailed pair*, an important circuit extensively used in balanced amplifiers. Each active device has a load resistor in its collector and shares a load resistor in its emitter. A signal in at the first stage causes signals out at both collectors and at the emitter, the ratio of the signal amplitudes being approximately *Rc*/*Re*.

The emitter signal is used to drive the second valve or transistor, whose base is held at a constant voltage. As we have seen when dealing with bias, a positive voltage at an emitter is equivalent to a negative voltage at a base providing the other electrode is held at a constant voltage, and the second stage then acts also as an amplifier with approximately the gain (*Rc*/*Re*) of the first. Because of this, the signals out at the two collectors are equal in amplitude, but the first is inverted in comparison with the input and the second, which has used the two stages, is not inverted (the signal at the emitter is not inverted, and the emitter input amplifier does not invert) so that a paraphase output is obtained. The outputs are of equal impedances, and some gain (approximately Rc/Re) is obtained.

A feature of this circuit is that any hum or other disturbance on the power supply line has no effect on the signals, because it causes the same rise or fall in each output. When the paraphase signals are recombined by inverting one and adding it to the other, one hum signal is now inverted and added so as to cancel out the other (Fig. 3.8(d)).

Note the bias arrangements usually necessary for this amplifier. Fig. 3.8(e) shows a circuit variously known as the *paraphase* or *see-saw*. A valve or transistor is connected with a feedback resistor from collector to base and an input resistor of equal value so that the gain is exactly unity but the output is inverted. This stage may be

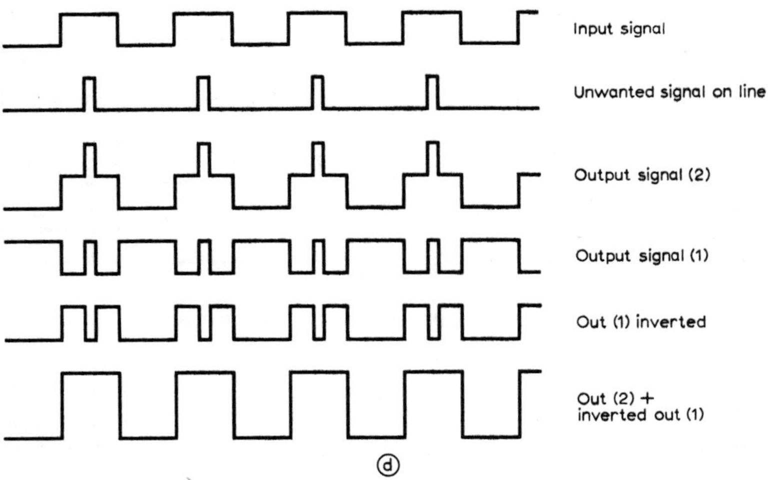

Fig. 3.8 (d) Paraphase circuits: showing how a long-tailed pair rejects unwanted signals in phase.

attached to any amplifier so that a normal and an inverted output are each obtainable.

The inverted output is of low impedance and can be used to drive low impedance loads and capacitive loads; it is frequently found in oscilloscope circuits driving one of a pair of deflector plates. Bias arrangements for this type of amplifier are straightforward. *Capacitive* see-saw circuits are possible and sometimes used.

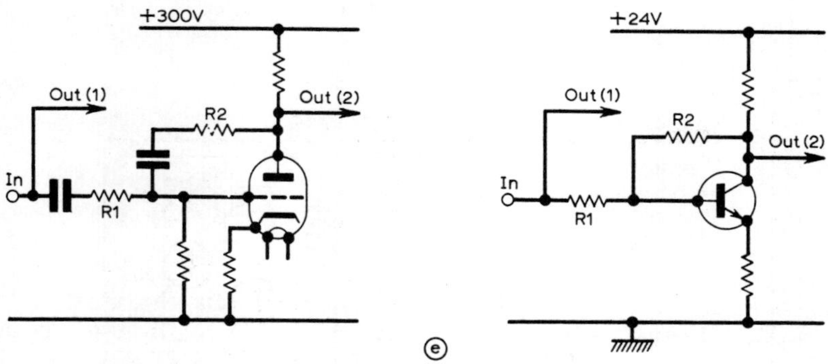

Fig. 3.8 (e) Paraphase circuits: see-saw circuit. For paraphase operation, R1 = R2.

Feedback in L.F. Amplifiers

Fig. 3.9 shows examples of feedback circuits used in l.f. amplifiers for various purposes. Fig. 3.9(a) shows a straightforward feedback loop used for the purpose of reducing internally generated noise and distortion, extending bandwidth and linearity, and stabilising the gain of an amplifier against variations due to changes in components (assuming the components in the feedback loop are stable).

Fig. 3.9(b) shows a feedback loop used to impose a definite shape

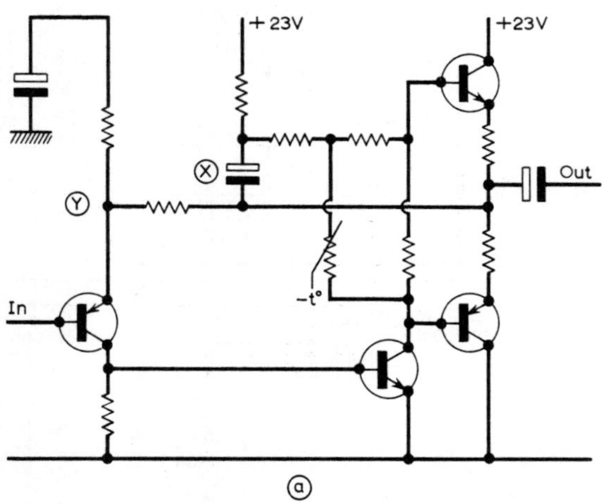

Fig. 3.9 Feedback loops. (a) negative feedback to point Y. Note also the bootstrapping connection to X. (Part circuit of the Philips N4407 stereo tape recorder.)

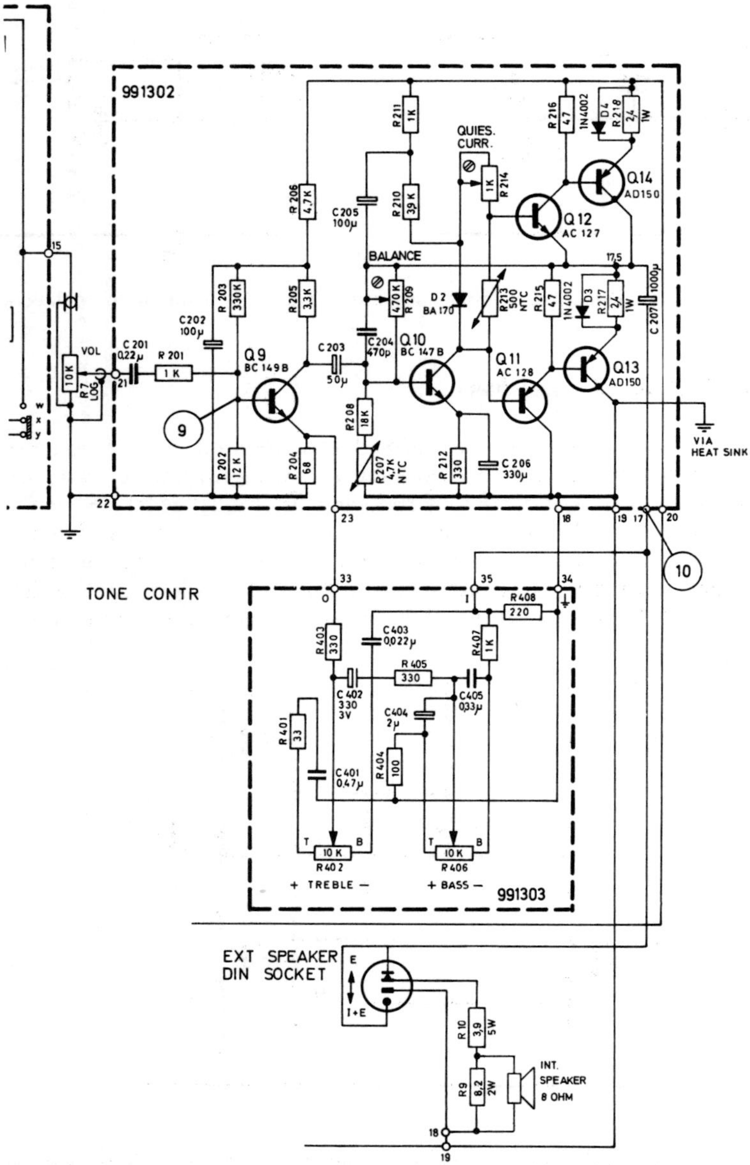

Fig. 3.9 (b) Feedback loops: treble and bass boost and cut through feedback network as used in the Tandberg 1700 Series tape recorders.

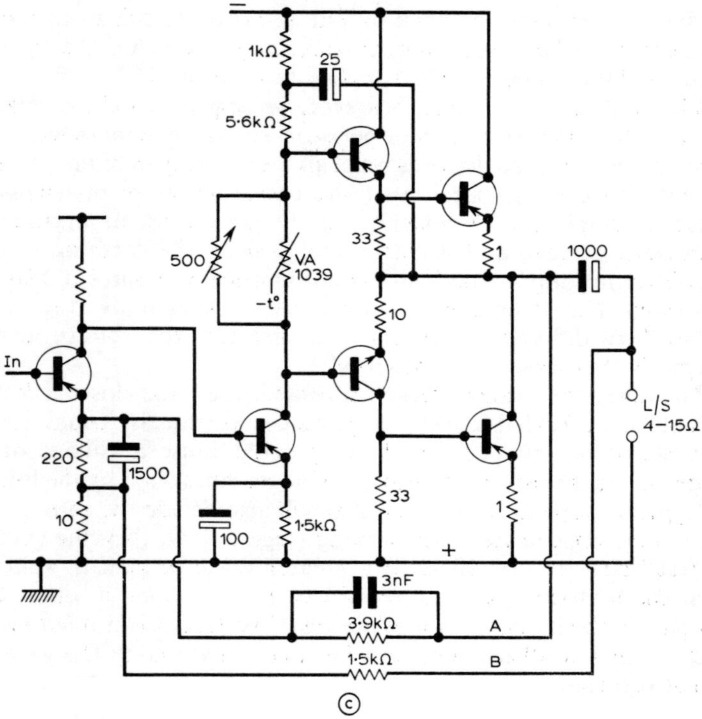

Fig. 3.9 (c) Feedback loops: portion of Leak amplifier, 1966 vintage, in which feedback loop A is added to correct phase-shift problems.

to the graph of gain against frequency, and 3.9(c) shows a feedback loop with a time-constant used to prevent oscillation which would have occurred otherwise due to a phase shift in the transistors. Fig. 3.9(d) shows a feedback loop intended to boost the amplifier's gain at one particular frequency at the expense of others (a *selective* amplifier).

Op-Amps as L.F. Amplifiers

When op-amps are to be used as l.f. amplifiers, the design problem consists of choosing biasing conditions and selecting the gain and bandwidth required. Because the bias requirements at the input of an op-amp are very small and the gain very large, an op-amp is practically always biased by some form of d.c. feedback arranged so that the resistance between output and input terminals provides the current for bias. Typically this value might be 3μA and a current of around this value must be supplied to *each* input.

The feedback bias must, however, be applied to the inverting input only, otherwise stable operation could be impossible. The value of resistor used depends much more on the gain required than on bias conditions, because of the construction of op-amps as balanced amplifiers. For example, in the normal use of an op-amp with both positive and negative supply lines, the earthing of the non-inverting input (marked +) automatically ensures a bias at this point. The other input (−) must now be held at a voltage only about 1mV different from this to ensure that the voltage at the output is also about earth potential.

The lower the value of feedback resistor used, the closer will this condition be held, but very large values of resistor could cause trouble, as the bias current flowing could cause a voltage drop large enough to make the output no longer controlled by the input. A 'typical' value is 100kΩ, and this will also decide the gain, for if the source impedance of the driving stage is 1kΩ, then the gain is 100kΩ/1kΩ = 100 or 40dB. If a greater value of gain is wanted, then the feedback path can consist of two resistors in series, decoupled at the junction so that the signal voltages are not fed back, and another feedback loop can be used for deciding the gain of signal voltages.

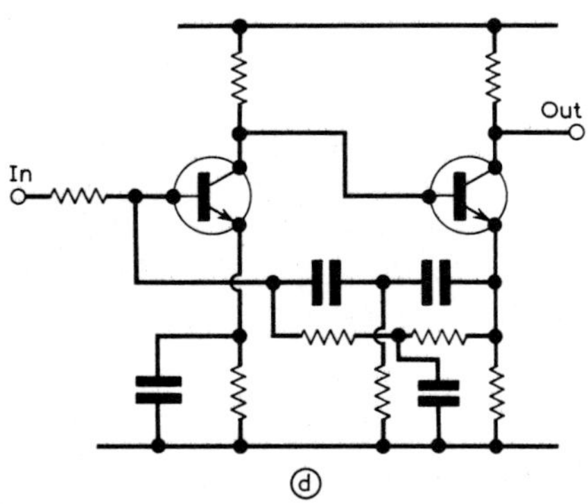

Fig. 3.9 (d) Feedback loops: selective frequency feedback.

Because the gain of the op-amp is decided by feedback, precautions have to be taken to ensure that the phase of the output compared to the input never shifts sufficiently to cause the feedback to become positive, or, if it does, to ensure that the gain is then

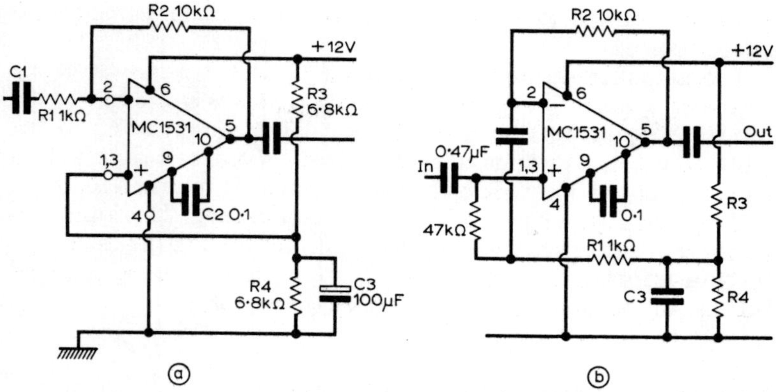

Fig. 3.10 Op-amp circuits: (a) inverting, (b) non-inverting. (*Motorola Ltd*).

insufficient to cause oscillation. The condition for stability is usually that the gain should decrease by 6dB for every doubling of frequency as the gain decreases to zero at high frequencies. This condition is often referred to as a 'slope of 6dB per octave at 0dB'; The use of the word octave (8-notes) for a doubling of frequency comes from music where traditionally (in the West, at least) there are 8 notes (counting inclusively) in a step from one frequency to double that frequency.

The usual way of ensuring this slope is to add a capacitor or capacitors to form a feedback loop inside the op-amp, connection points being provided for this purpose. The capacitors *could* be added at later stages, but the greater signals present would require greater currents to charge and discharge the capacitor, affecting the slew rate of the amplifier (its ability to handle large amplitude signals with the same frequency response).

Values of around 0·05μF are often used, especially if the frequency response is deliberately to be restricted, but lower values can be used if more care is taken or in other compensating networks. Manufacturers of ICs supply graphs showing the gain and bandwidths available with different compensation networks, and it is advisable to adhere closely to these values.

A typical op-amp l.f. amplifier is shown in Fig. 3.10(a). In this case a single supply line has been used, and one input (the non-inverting one) has been returned to a mid-voltage, well decoupled, leaving the bias network to supply the inverting input. The gain required is 20dB, and this is supplied by the MC1531 IC by using a bias resistor of 10kΩ and a source resistor of 1kΩ. Stability is ensured by using a 0·1μF capacitor as compensation between pins

9 and 10, and the bandwidth to the point where gain is reduced is about 100kHz.

The amplifier shown is phase-inverting and if a non-inverting amplifier is wanted (Fig. 3.10(b)) then the bias is still taken to the inverting input, but the non-inverting input is connected to the mid-voltage line through a load resistor. When two supply lines (equal + and −) are used, then the + input is returned to earth, making the components *R*3, *R*4, *C*3 unnecessary. The input impedance of a non-inverting amplifier can be raised in the usual way by bootstrapping, and very high values (1—10MΩ) can be obtained because of the large values of feedback (loop *gain*).

Pulse Width Modulation ('Class D') Amplifiers

Imagine a switch, as shown in Fig. 3.11, which can be switched to either a positive or a negative voltage. If the switch is changed over and back every 0·5μS the output obtainable will depend on how much time is spent in each position. If equal times are spent at both positive and negative positions, the output will be a square-wave at 1MHz, and a d.c. meter would register no deflection.

If, however, the switch spends 90% of its time (0·9μS) in the positive position and only 10% of its time (0·1μS) in the negative position the meter would register a deflection showing a positive output; conversely, if the switch spends 90% of its time in the negative position the meter would read a negative current.

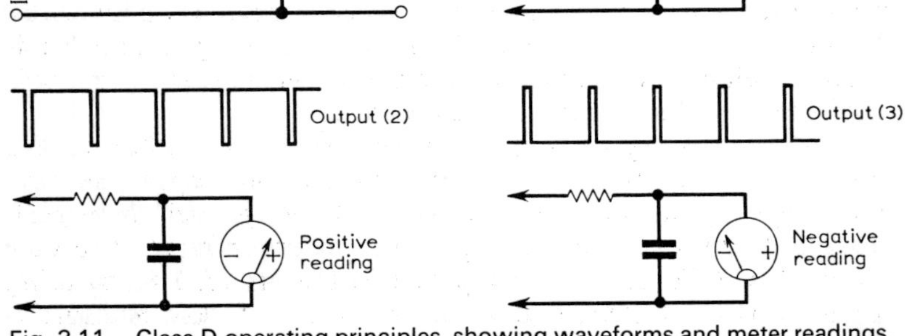

Fig. 3.11 Class D operating principles, showing waveforms and meter readings.

The attractive feature of this scheme is that the switch is not dissipating any power, since it only acts as a connection; if we make the switch from a switching transistor, and we cause the time spent in each position to vary according to a signal voltage, we can obtain very large amplifications and high power outputs with very little dissipation.

This type of circuit is known as PWM (pulse width modulation), and has been used for amplifiers of very high power output and very good linearity. Such amplifiers, to give good results, must be very carefully designed and required elaborate pulse generating and modulating circuits; simple versions are possible, but do not have the advantages of the elaborate circuits.

CHAPTER FOUR

TUNED AMPLIFIERS

At low frequencies, it is possible to make amplifiers using simple techniques which amplify from nearly d.c. (z.f.) up to 20kHz or so. If we wish to amplify signals which are arriving at a frequency of 1MHz, and which have a bandwidth of only 20kHz on each side of 1MHz, it would be very wasteful to build an amplifier capable of amplifying from d.c. up to 1·020MHz (though this is possible) because the frequencies below 0·980MHz (1MHz less 20kHz) would never be passed through the amplifier.

In all amplifiers, the figure (Gain × Bandwidth) is fixed by the devices we use and the higher the Gain × Bandwidth figure the more costly the amplifier. High frequency amplifiers, therefore, are made to amplify a limited bandwidth around a centre frequency rather than to amplify all frequencies up to this centre frequency, since the latter would require a very large bandwidth, and so the gain would be very low unless a very complex amplifier were used. This complexity is, however, possible with integrated circuits.

Amplifying a Limited Bandwidth

Amplification of a limited bandwidth around a centre frequency is achieved by using a tuned amplifier whose load impedances are tuned circuits. Fig. 4.1 shows a simple tuned amplifier with a parallel-tuned circuit as its load. The gain of this amplifier is at a maximum when the impedance of the load is at a maximum, and this occurs when the signal frequency equals the frequency of resonance. We would therefore expect the graph of gain against frequency to look rather as shown in Fig. 4.1(c), where the maximum gain is at the resonant frequency of the tuned circuit and the gain falls off rapidly as the frequency is changed from this resonant frequency.

Several stages of the simple amplifier of Fig. 4.1 could be coupled

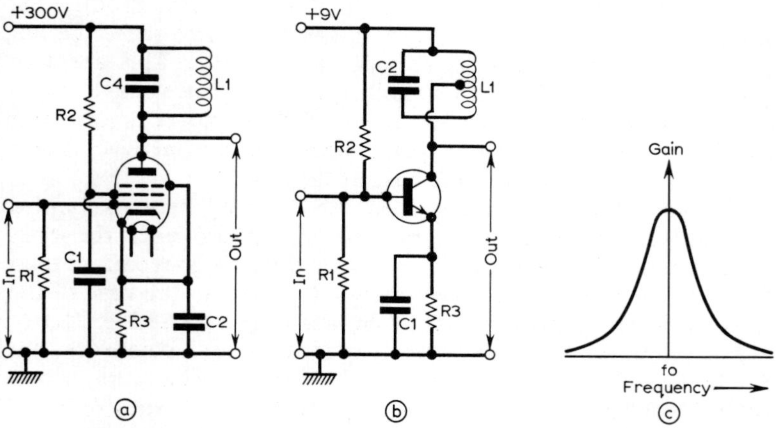

Fig. 4.1 Using a tuned circuit as a load impedance: (a) valve circuit; (b) transistor circuit. Note the use of a tapped inductor to match the comparatively low impedance of the transistor; (c) Gain/frequency graph for the amplifiers shown.

together to give an amplifier of higher gain and of narrower bandwidth. The gain in dB for each stage can be added up to give the total gain (allowing for coupling losses) and the total bandwidth is the average bandwidth for one stage divided by the number of stages.

Coupling Stages Together

The systems of coupling stages together which are used on low frequency untuned amplifiers are not very useful when used with tuned amplifiers. At lower frequencies, resistor-capacitor coupling can be used, but there will be stray capacitances across the resistor load of each stage, and they will cause the gain to be 3dB down in each stage at the frequency given by $f = 1/2\pi R.C$ where R is in MΩ and C is in pF. For example, if there is stray capacitance of 10pF across 10kΩ (1/100MΩ), then the 3dB frequency is about 1·6MHz. In addition, this method of coupling gives no tuning effect, though the bandwidth can be very large; it is used to some extent in integrated amplifiers and in untuned wideband amplifiers and will be examined again in Chapter 5.

If the coupling method is to transfer signals of a chosen frequency, as is intended, then the load must be a tuned circuit. One method is to use either a series or parallel tuned circuit in the coupling.

When a series tuned circuit is to be used, the aim is to produce the maximum signal *current* in the circuit at resonance, since the impedance of a series tuned circuit is least at resonance. This is seldom useful in valve or MOSFET amplifier stages, where voltage amplification is more important, but can be usefully employed in transistor amplifiers, where a current drive can be arranged from the collector circuit of one transistor to the base of the next.

The series tuned circuit is then the coupling network, and a separate load must be used in the collector. At resonance, the series tuned circuit offers a much lower obstruction to signal current than the load, so routing the collector signal current to the base of the next transistor. Stray capacitance between the base and emitter of the driven transistor can be used as part of the tuning network, but at the expense of taking some of the current. This approach is often found in high-frequency tuned circuits, as for example in TV Band IV and V tuners.

When a parallel tuned circuit is used, it is used as a load for the driving stage, and reaches maximum impedance at resonance, so transferring the maximum signal voltage or current to the next stage. The impedance presented by such a circuit at resonance is known as its dynamic resistance, and is given by: L/CR. This will be in ohms if the units of L are μH and the units of C are μF, with ohms used for the units of R. (R in this case is the resistance in the parallel circuit). This formula assumes that the value of R is small compared with the dynamic resistance, as is usually the case in narrow bandwidth tuned amplifiers.

The bandwidth of such an amplifier, to the 3dB point from the central frequency, is f_0/Q, where f_0 is the frequency of resonance, the centre frequency, and Q is the constant defined by $2f_0L/R$ or $1/(2f_0CR)$ where R is again the resistance of the tuned circuit. Beyond the 3dB points, the response falls off at a rate of 6dB per octave (an octave is a doubling or halving of frequency), and the phase shift, which is 45° at the 3dB points, becomes about 80° at the 10dB point and approaches 90° at frequencies well away from resonance.

The use of single tuned circuits of these types is fairly simple and safe, in the sense that the phase shift is comparatively small and gradual, so that small amounts of feedback from the output of a stage to its input are not too serious. The selectivity of such circuits, however, is not very high, so that several stages may have to be used to obtain a desired selectivity. Greater selectivity can be obtained by coupled tuned circuits and, oddly enough, such circuits can also be put to use to obtain more bandwidth (though not at the same time!).

A large number of methods of coupling two tuned circuits exist, including common impedance coupling where the two circuits are connected across a common capacitor or inductor to earth, but the most common coupling is transformer coupling, where the two inductors are wound close enough together to ensure some mutual inductance. The closer the coupling, which depends on the magnetic circuit as well as on the physical closeness of the coils, the more efficient is the transfer of the signal and the greater the mutual inductance. The closeness of coupling is measured by a coupling coefficient (k), which, for transformer coupling, is given by $M/\sqrt{(L1.L2)}$, where M is the mutual inductance between the coils whose self-inductance values are $L1$ and $L2$. The same units of inductance must be used for all three quantities.

The response of the tuned coupling now depends on the value of k, the coupling coefficient, and Q, the Q-factor of the coils (assumed equal, otherwise the average $\sqrt{Q1.Q2}$ can be taken). If coupling is very loose, with the value of k very small, then the bandwidth of the circuit is reduced to about 0·6 of the value for a single tuned circuit. Note, however, that with such loose coupling, the amplitude of signal transferred at resonance may be smaller.

At higher values of coupling, the bandwidth to the 3dB point is greater than for a single tuned circuit, but the rate of fall beyond the 3dB points is greater, 12dB per octave, so that the circuit is more selective at frequencies beyond the 3dB points. At values of k^2Q^2 greater than 2, the graph of response against frequency starts to show a rise at the frequency at which a single tuned circuit would be *down* by 3dB, and this peak becomes greater if greater values of k^2Q^2 are used.

For a value of k^2Q^2 equal to 4, each peak is 2dB above the response at the centre frequency of resonance. The peaks produced by 'over-coupling', as this effect is termed, can be smoothed down by loading the circuits with resistors in parallel, so producing a useful wide-band response (but, of course, with less gain, since the product G-B is preserved).

The phase-shift produced by coupled circuits is, as might be expected, greater than that of a single circuit. For loose coupling with k almost zero, the phase shift is 60° at the 3dB points and 90° at the 6dB points, but the more tightly coupled circuits produce smaller phase shifts at the same frequency away from resonance. The maximum phase shift at a frequency well away from the resonant frequency is 180°, so that a trace of feedback from the output to the base of the driving transistor can cause trouble with instability.

In early types of transistors, the feedback path from collector to base was equivalent to several pF of capacitance in parallel with a fairly low resistance, and this was sufficient to cause instability in amplifiers using a double tuned transformer load unless balanced out by a process called unilateralisation (when the feedback was balanced out at all frequencies) or neutralisation (when the cancellation applied only over a limited range). Both of these methods depended on adding networks which counteracted the positive feedback with some negative feedback.

For most purposes, the feedback in transistors of modern design is so small as to make such precautions unnecessary, but unilaterisation networks, consisting of series *RC* connected from one collector to the collector of the previous stage may sometimes be found in older circuits or where transistors of modern design are used near their frequency limits.

Features of Amplifiers of Different Design

The *470kHz narrow bandwidth* amplifier of Fig. 4.2, typical of those found in older transistor radios, uses two stages with tuned transformer coupling, giving high gain and narrow bandwidth. Since low-cost (at the time) transistors are used, working near their frequency limits, some neutralisation is needed, and is provided by the networks shown. Any changes in the components used for neutralisation or the position of the wiring may cause oscillation; a change of transistors to types with different internal feedback may also cause oscillation because the neutralising networks will no longer be matched to the transistors.

The *10MHz medium bandwidth* design of Fig. 4.3 uses double

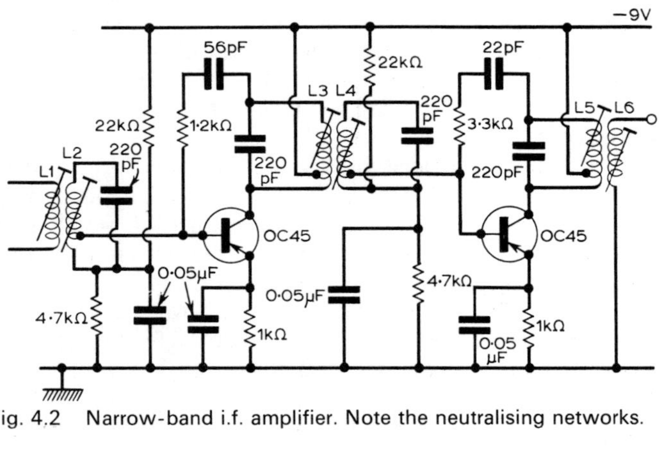

Fig. 4.2 Narrow-band i.f. amplifier. Note the neutralising networks.

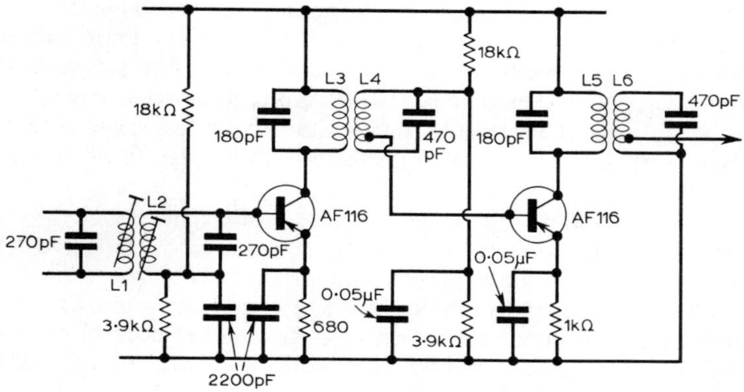

Fig. 4.3 (above) Un-neutralised i.f. amplifier.

Fig. 4.4 (below) 60MHz wideband valve amplifier. (*Decca Radar Ltd.*)

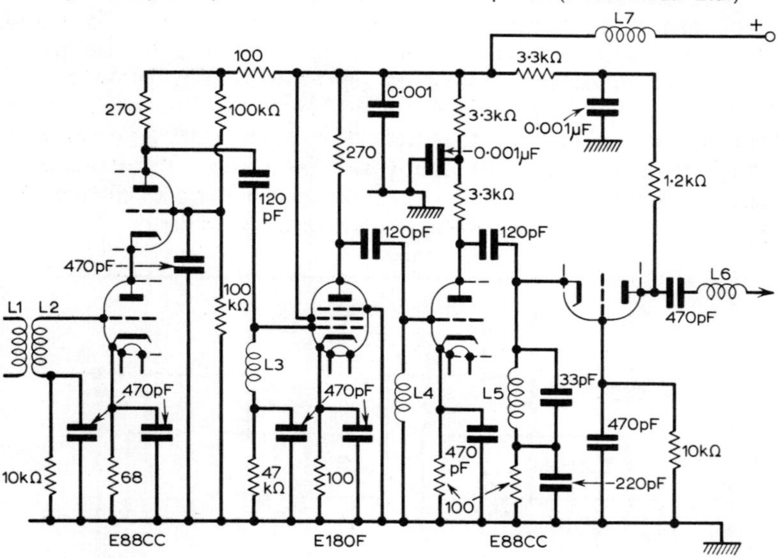

tuned transformers with no neutralisation, due to the use of more advanced types of transistors working well within their frequency limits. Note that the secondary of the coupling transformer is tapped so that a low impedance lead can be taken to the base of the next transistor, but that the tuning capacitor is across the whole of the winding. The collector return line is earthed to make decoupling easier.

In the *60MHz wideband valve amplifier* of Fig. 4.4 the input is fed into *V*1 grid by a tuned transformer, the secondary winding being tuned by the grid-cathode capacitance of the valve. The first stage, *V*1a and *V*1b is a cascode amplifier, a common-cathode triode coupled directly to a common-grid triode. The two portions of a cascode amplifier together produce characteristics very similar to those of a single pentode but produce much less noise, and for this reason are very often preferred for the early stages of v.h.f. amplifiers.

The second stage is a pentode whose grid circuit is tuned by an inductor *L*1, and the grid-to-cathode capacitance of the valve. The output of the pentode is passed to another cascode pair, this time coupled by a capacitor. Note that the circuit diagram has been simplified, omitting details of the circuitry used to control the gain and bandwidth of the amplifier. The working bandwidth is 16MHz, typical of the bandwidth used in radar receivers.

The *60MHz wideband transistor amplifier* of Fig. 4.5 is transformer coupled and neutralised; the primary windings of each transformer are shunted by a damping resistor to lower the *Q* and so give greater bandwidth. The emitter resistor is in two parts, one undecoupled as a precaution against instability. Note the use of beads of Ferroxcube which are threaded on to the wire leads at various points. This has the effect of increasing the inductance of that portion of wire, and greatly aids decoupling. Particular care should be taken in servicing such an amplifier not to disturb the position of these beads, as this could cause instability.

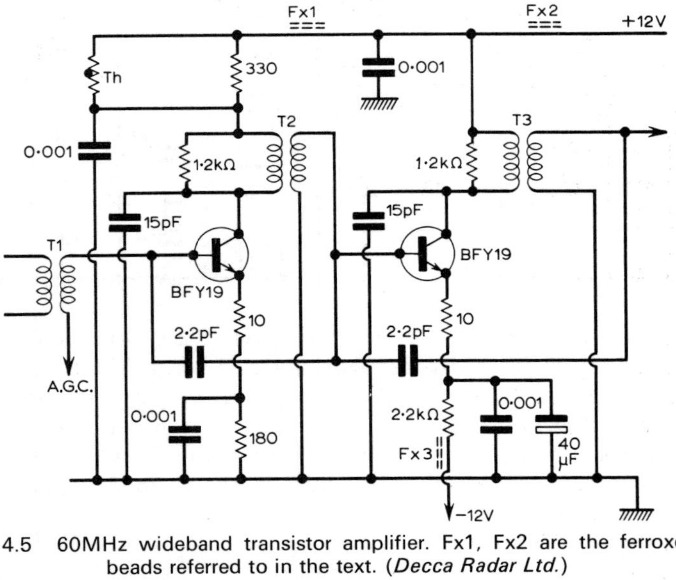

Fig. 4.5 60MHz wideband transistor amplifier. Fx1, Fx2 are the ferroxcube beads referred to in the text. (*Decca Radar Ltd.*)

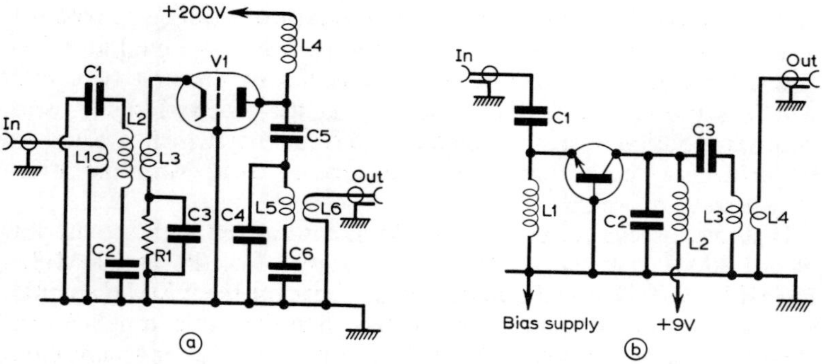

Fig. 4.6 Common-grid (a) and common-base (b) circuits.

Higher frequency amplifiers. At frequencies of 100MHz to around 1000MHz, triodes and transistors of suitable design (particularly as regards low capacitance between electrodes) may be used in common-grid or common-base circuits of the type shown in Fig. 4.6. Tuning may be by small loops of wire, parallel lines or portions of waveguide depending on the frequency. In such amplifiers, the physical form is more important than the circuit diagram, and small disturbances to the arrangement of components, wires or screens can make a large difference to the operation of the circuits.

Frequency Changing

When the received radio signal (r.f.) is too high for existing devices to provide the required degree of amplification, or when the need to operate at different frequencies would require changing the time constants in a large number of stages, the technique of frequency changing is used. Fig. 4.7 shows in outline the methods used.

The received signal is fed into a stage where it is mixed with an

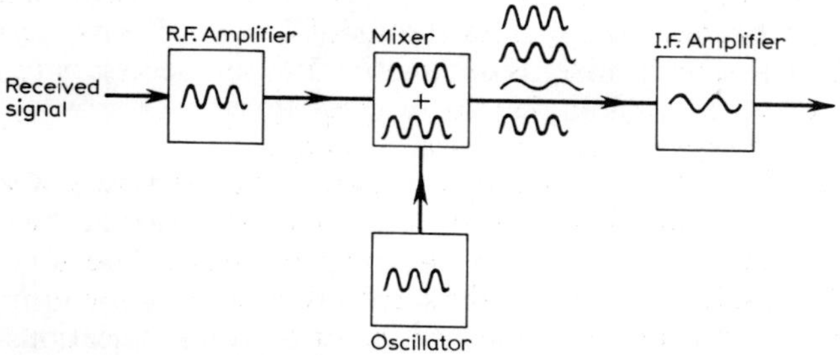

Fig. 4.7 Block diagram of frequency changing operation.

unmodulated sinewave. The mixing operation produces a total of four outputs—the two signals which were fed in, a signal at a frequency which is the *sum* of the frequencies of the first two, and which is modulated in the same way as the received signal, and another modulated signal which is at a frequency which is equal to the *difference* between the signal frequencies fed in. An illustration may help to make this clearer.

Imagine a signal received at 200MHz and mixed with the output of a 170MHz oscillator. The output signals will be at 200MHz, 170MHz, 370MHz and 30MHz, and all except the 170MHz signal will be modulated. Since the 30MHz signal is at a much lower frequency than the others and is therefore much easier to amplify, further amplification can be carried out at this frequency. This method of dealing with received signals at high frequencies is called superheterodyne reception, the word being an amalgam of *supersonic heterodyne* (the mixing is the heterodyne part and the frequency obtained is supersonic, meaning in this case above the frequency of sound).

The frequency used for mixing with the incoming signal is generated by an oscillator (the local oscillator) which forms a most important part of the receiver and which must maintain high stability of oscillating frequency if constant retuning is not to be needed.

Features of Superhet Working

(1) When variable tuning is needed, only the local oscillator and the r.f. stages need to be tuned, so long as the difference between the frequencies is kept constant.

(2) The difference frequency or intermediate frequency (i.f.) can be amplified to a much greater extent than would be possible if each stage had to be variably tuned.

(3) It is always possible to generate oscillations at any frequency which can be received, so that the amplification of any signal is possible if a suitable mixer is obtainable. It is not necessary to have devices capable of amplifying the received signal.

Frequency changing may be carried out by:

(a) A valve with several grids (heptode or hexode), two of which act as control grids. This was the technique most used at the lower radio frequencies in the days of valve receivers. The oscillator could be separate, or part of the same valve; usually the former in 'domestic' radio receivers and the latter in communications-type receivers.

(b) A valve or transistor in which signals may be injected at grid/base and at cathode/emitter. Usually signals are taken into an oscillating transistoı which is oscillating because of a feedback between collector and emitter, so that the signals can be taken to the base. This method (self-oscillating mixer) is particularly popular in transistor radios for medium wave working.

(c) A diode into which the two signals are fed. In this case, the oscillator must be separate. This technique is almost universal at very high frequencies. The use of a transistor instead of a diode does not change this mode of operation as long as both signals are fed into the same electrode.

(d) A bridge circuit, which may also be used for modulation or demodulation in other applications. In the past, bridge methods have not been widely used outside specialised communications equipment, but the advent of low-cost integrated circuits has made their use more common. An example of a bridge circuit is the ring modulator circuit whose arrangement is shown in Fig. 4.8, clearly demonstrating the derivation from the bridge circuit; using the idea that the ratio of each pair of arms is controlled by an incoming signal.

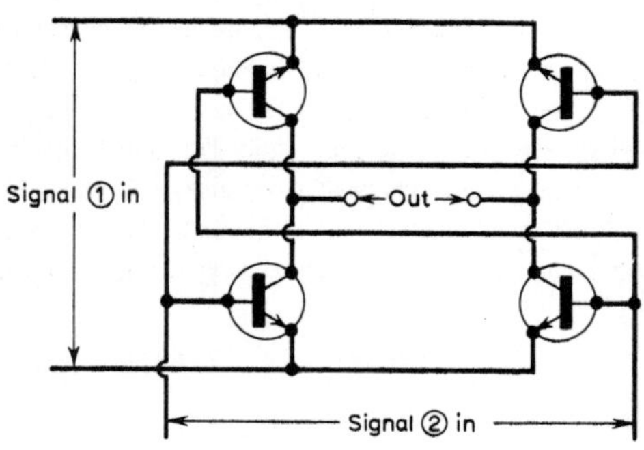

Fig. 4.8 Bridge mixer/modulator.

The application of one signal to the bases switches one pair of transistors on and the other pair off, so that the amplitude of the output signal depends on the resistance between collector and emitter of the transistor in the 'on' condition. As the control signal changes in voltage, the transistor resistance changes, so changing the amplitude of the signal passed. Because of the balanced nature of the circuit, dealing with both polarities of signal, there is very little 'leakage' of unwanted frequencies.

In integrated receiver circuits, the balanced amplifier is used as a mixer, with the signal voltage applied to one base of the balanced pair and the oscillator voltage to the base of the other. As the output from the collector of such a circuit depends on the difference between the two base currents, the frequency obtained is the desired difference frequency, along with the usual others. This again is an example of a circuit which could be expensive to use with separate components but which is economically feasible in integrated form.

R.F. Amplifier

Mixing is a noisy operation, and if the mixer is the first stage of an amplifier, the sensitivity of the amplifier will be limited by the noise generated in the mixer. For this reason it is usual, except in domestic radios working on the Medium and Long wavebands, to use a stage of amplification before the mixer. This stage should have the highest possible gain and the lowest possible noise level. It need not necessarily be tuned, but since the noise of a stage increases with its bandwidth, it is desirable that the r.f. stage, as it is called, should be tuned.

Up to 150MHz, this r.f. stage might be a pentode, up to about 30MHz a transistor operating in the common-emitter connection. At frequencies above 30MHz, common-base transistor circuits are more used; cascode transistor and valve circuits can also be found.

At frequencies of 500MHz upward, common-base connection of some germanium transistors can still be used, and u.h.f. triodes can be used in the common-grid connection, with tuning by lines or cavities. Above 2,000MHz (2GHz), travelling-wave tubes, amplifier klystrons and other microwave devices must be used.

A.G.C.

In all transmitted signals, unless the receiver is very close to the transmitter, fluctuations of signal strength can cause difficulty. The effect of fading is caused by 'interference', which is the beating together at the receiver of signals which have arrived by different paths. This is because the layers in the upper atmosphere reflect radio waves to an extent which depends on the nature of the layers (their weather) and the frequency of the signals. These layers become ionised by charged particles streaming from the sun and are constantly varying, and so their ability to reflect signals is also constantly varying.

The net result is that some signals reaching the receiver may have come in a straight line, some may have been reflected once from a low (60 miles high) layer of the *ionosphere* and some may have reflected from a high layer. Multiple reflection between the Earth and the Ionosphere is also possible. At short wavelengths, reflections from aircraft also cause severe fluctuations in signal strength.

This problem is solved at the receiver by automatic gain control, or a.g.c., which involves using a feedback loop to change the gain of an amplifier in an attempt to keep the output as constant as possible. The feedback is not at signal frequency, as this would create stability problems, but at d.c. The carrier, rectified at the detector, is used to provide a bias voltage for earlier stages in the amplifier, so controlling their gain.

With valves, this was accomplished with 'variable mu' valves, whose grid windings were of progressively decreasing pitch. This caused different portions of the valve to cut off at different grid bias voltages, causing the gain to be less at large negative bias, because the portion of the grid being used had a wider grid-wire spacing. The behaviour of transistors is more complex, and control is possible either by reverse bias (reverse a.g.c.) or, in suitable designed transistors, by forward bias (forward a.g.c.).

For television signals, the simple process of using the rectified carrier to provide an a.g.c. signal is unsuitable, because the carrier signal is suppressed at the transmitter, leaving the upper sideband and a portion of the lower one. Rectifying the video signal is unsuitable (though widely used) because the amplitude of the video signal varies with the content of the picture as well as with the r.f. amplitude. Rectification of the sound carrier is unsuitable because there is no guarantee that vision and sound signals will fade together.

The most certain method is to sample the voltage of the video signal at a time when there is no modulation (black level) of *video*, though there is still an r.f. signal being transmitted. This could be done with some difficulty on the earlier 405 line system, but can be more easily achieved on the 625 system, where the tips of the sync pulses represent full carrier amplitude, and so can be rectified to provide a.g.c. (Television a.g.c. systems are described in detail in *TV Technician's Bench Manual*, published by Fountain Press.)

R.F. Transmission Techniques

If transmission consisted of nothing more than radiating a radio frequency wave, a transmitter would consist of nothing more than an oscillator coupled to an aerial. For some purposes, this might be sufficient; for example, if a power supply had to be monitored at a distance, the presence of a signal would signify power on and the absence of a signal power off, and no more elaborate equipment would be needed.

For most purposes, however, the transmission of information in the form of code, audio frequency signals or video frequency signals is required, and so a transmitter consists of three basic portions:

(1) The oscillator and frequency-multiplier provides the carrier frequency at low power which may be amplified directly or multi-

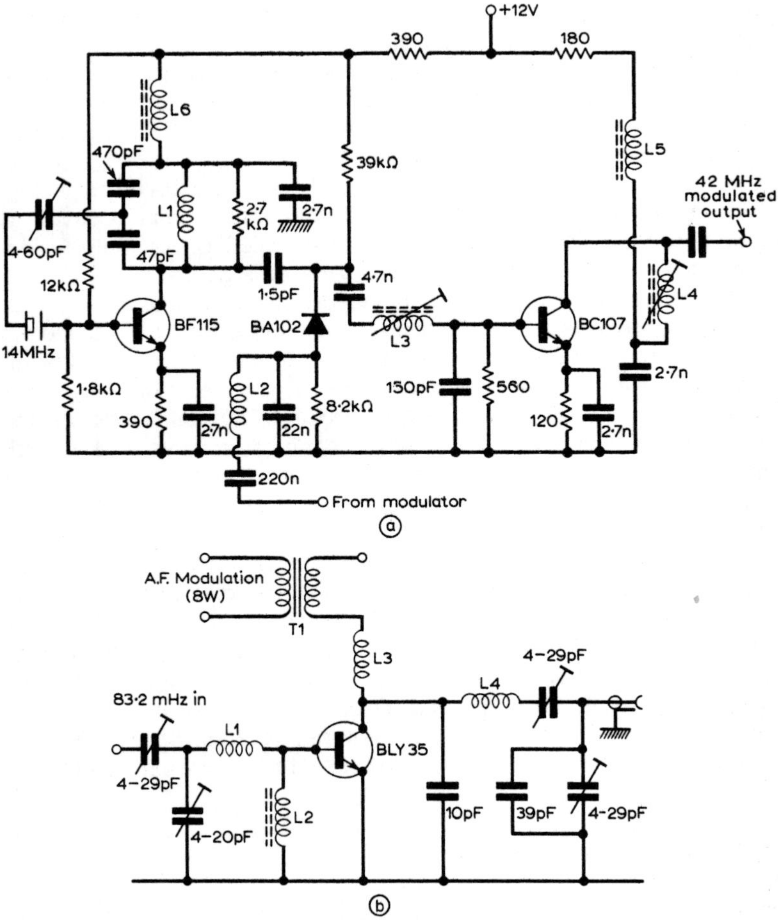

Fig. 4.9 (a) frequency modulated oscillator; (b) amplitude modulated P.A. stage. (*courtesy of Mullard Ltd.*)

plied in frequency. The reason for using multipliers is that it is much easier to make a stable oscillator if a low working frequency is chosen. The generated frequency can then be doubled, tripled etc. up to any desired multiple to give the frequency required for transmission.

Multiplication is carried out in deliberately under- or over-biased stages, so that the sinewave signal in is distorted to a squarer shape in which a large number of harmonics is present. The harmonic (multiple) required is selected by using, as the load of the stage, a circuit tuned to the frequency of this harmonic. Losses are high, as the percentage of the higher harmonics in the output signal is small. In this way, crystal controlled oscillators can be used to supply outputs at frequencies which are too high to be generated by the crystals themselves.

(2) The modulator stage is the stage which mixes the information (code, audio, video etc.) into the radio frequency. Information is in the form of a varying voltage which may be used to vary the amplitude of the r.f. (amplitude modulation, a.m.) or its frequency (frequency modulation, f.m.) or its phase (phase modulation, p.m.), or some combination. Modulation will generally be carried out in balanced (ring or bridge) modulators similar in design to that discussed under the heading of mixers.

(3) The power amplifier (PA) stage is used to amplify the output power of the oscillator, frequency multiplier or modulator to the power level required for transmission, and to feed this power to the aerial, using a matching transformer if the output impedance of the PA differs from the radiating resistance of the aerial. In some cases, modulation may be carried out at the PA stage.

A.M. and F.M.

Amplitude modulation is usually carried out at the power amplifier or as close to it as possible. The reason for this is that amplification of an amplitude modulated waveform must be Class A or B to avoid excessive distortion, but power amplifiers preferably work in Class C, where the valve or transistor is switched on for only a short time in each cycle, the rest of the cycle being provided by the resonant circuit, known as the 'tank', connected to the output.

Frequency modulation, on the other hand, is preferably carried out at the oscillator so that the generated frequency can be varied in proportion to the voltage used for modulation. Fig. 4.9 shows two simplified modulator circuits, one for f.m. and one for a.m. The f.m. modulator depends on the fact that a back-biased semi-

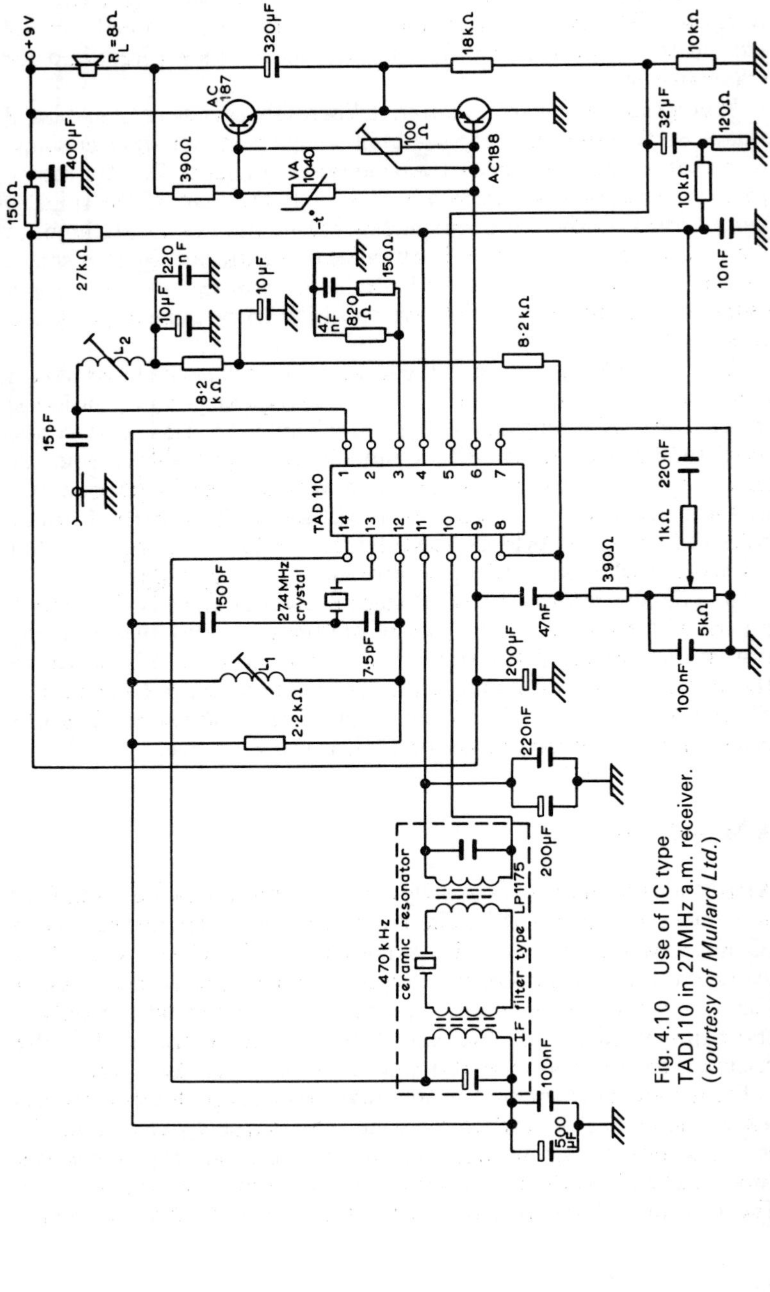

Fig. 4.10 Use of IC type TAD110 in 27MHz a.m. receiver. (*courtesy of Mullard Ltd.*)

conductor diode has a capacitance which varies with the voltage across it. If this capacitance forms part of a tuned circuit the frequency of resonance will vary with the voltage across the diode.

If an audio signal is fed to the diode, the oscillator output varies in frequency as the audio signal varies in voltage; frequency modulation has been achieved. This frequency modulated output can then be frequency multiplied and amplified without risk of distorting the modulation because this depends on frequency shift and not on amplitude.

The a.m. modulator operates on the PA stage, and uses the audio signal to vary the collector voltage applied to the PA transistor, so that the amplitude of the output must correspond to the amplitude of the audio signal. This method of modulation is more suitable for lower power PA stages, since the audio signal must be capable of supplying the current demands of the PA stage; in other words, the power output of audio must equal the power supply to the PA stage. In high power PA stages, modulation is applied to the signal at the PA; in a valve stage this would mean modulating the grid voltage of a linear power amplifier.

Integrated Circuits in Tuned Amplifiers

Several integrated circuits are now available for use in tuned amplifier applications. As inductors cannot be included within the integrated circuit, and capacitors are comparatively difficult to form within close limits, the amplifiers consist of the usual balanced

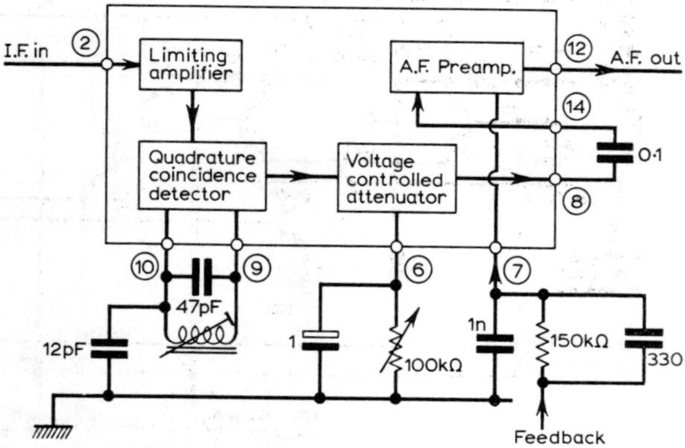

Fig. 4.11 Use of IC for sound detection and preamplification in Ferguson Colourstar 17″ TV. The i.f. input is from a crystal resonator and the output is to an output stage from which feedback is applied to pin 7.

design, with the tuning components applied externally. In addition, other circuit functions such as f.m. limiting and detection or a.m. detection and some audio preamplification can be included in such a circuit. Several circuits for a.m. and f.m. radio i.f. circuits exist, along with integrated circuits which carry out practically all of the functions of a medium wave radio set. In addition, TV-range i.f. amplifiers are also available.

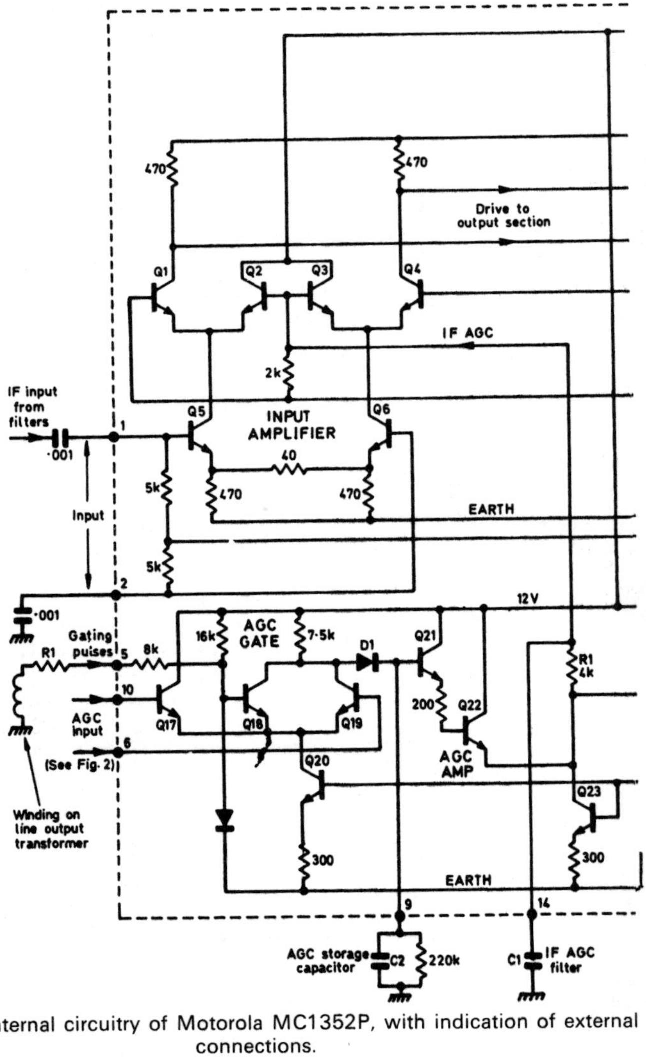

Fig. 4.12 Internal circuitry of Motorola MC1352P, with indication of external connections.

As an example of the use of an IC in an a.m. receiver circuit, Fig. 4.10 shows the Mullard TAD110 used as mixer-oscillator, i.f. amplifier, detector and audio preamp in a receiver operating at 27MHz. This indicates the frequency capabilities and range of functions possible in such ICs.

The use of an integrated circuit in a colour TV receiver is indicated in Fig. 4.11, showing the circuit used for sound detection and pre-

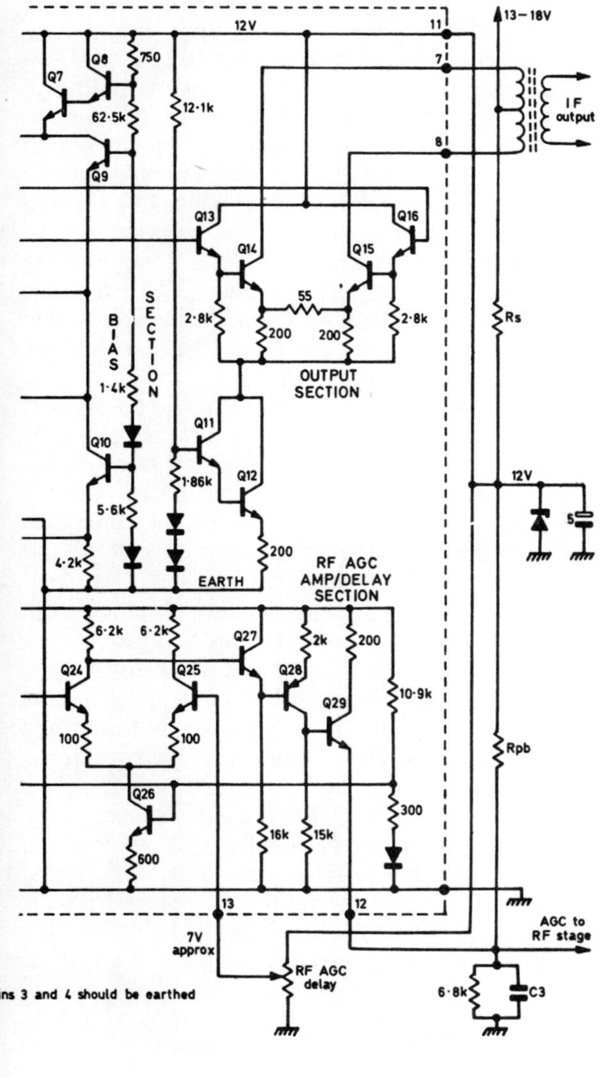

amplification in the B.R.C. Series 8000 receivers. The sound signal is picked off from the video detector as a beat signal between the sound carriers and the vision carriers. This beat signal is at the difference frequency of 6MHz, and is frequency modulated by the sound signal. This technique, almost universally used, is known as 'intercarrier sound'.

The sound intercarrier of 6MHz, with sidebands, is filtered to remove any chrominance signal and applied to the intercarrier amplifier. The IC used in this receiver amplifies, limits and detects the intercarrier signal, giving an output of about 100mV peak-to-peak. This is then amplified to about 450mV p-p by an audio pre-amplifier on the same chip, and so delivered to the output stage.

As an example of the internal circuitry used on an integrated circuit for i.f. use, Fig. 4.12 shows the internal circuitry of the Motorola MC1352P, which is a vision i.f. amplifier with a gated a.g.c. section for gain control. Because tuned circuits cannot be built in to the IC, the tuning is provided by external filters, feeding the 45MHz i.f. signal in at the input pins 1 and 2. This input is applied to transistors *Q*5 and *Q*6 which are connected in a balanced circuit designed to maintain constant emitter current so that the input impedance is unchanged by a.g.c. action.

The outputs of *Q*5 and *Q*6 drive the emitters of the compound pairs *Q*1, *Q*2 and *Q*3, *Q*4. The a.g.c. control potential is applied to the bases of *Q*2 and *Q*3 and acts so as to 'steer' signal into *Q*2, *Q*3 or into *Q*1, *Q*4, from which the signal is taken to the bases of *Q*13 and *Q*16. These latter transistors form the first part of a compound pair, and the output signal is taken from the collectors of *Q*14 and *Q*16 to be applied to an external i.f. transformer.

The a.g.c. section requires a gating pulse, usually obtained from the line output stage, a d.c. reference or control signal, and a video signal. The capacitor *C*2 acts as a storage capacitor for charge delivered through *D*1 and drained by *Q*21. Charge is delivered when the gating pulse arrives at pin 5, and the amount switched into the capacitor depends on the level of voltage at the emitters of *Q*17, 18, 19 at the time of the pulse.

As *Q*17 is supplied with video signal from the video detector, the level of *Q*17 emitter at the time of the gate pulse is the level of the line sync pulse, which is always proportional to r.f./i.f. signal amplitude. The voltage across the capacitor is applied to *Q*21, *Q*22 and taken through the filtering resistor *R*1, to the variable gain stage in the i.f. amplifier section and to the r.f. a.g.c. amplifier section.

It is then compared in the differential amplifier *Q*24, *Q*25 to a fixed reference voltage connected to pin 13. Stages *Q*27 to *Q*29 amplify the difference signal, so that the r.f. a.g.c. voltage changes from zero to ±7V for a very small change in i.f. a.g.c. voltage.

Crystal Filters

When one signal has to be selected from a large number of others at frequencies very close to the wanted signal, a receiver of very narrow bandwidth is required. A superhet receiver can be made to have a fairly narrow bandwidth, but for very narrow bandwidths (as for code reception) *crystal filters* can be used in the i.f. stages.

Single Sideband

When a graph of power output against frequency is plotted for an a.m. transmitter, the result is as Fig. 4.13. A large amount of power is radiated at the *carrier frequency* (the unmodulated radio frequency) and modulated power is given off at frequencies equal to carrier frequency *plus* audio (or video) frequency and also at carrier trequency *minus* audio (or video) frequency.

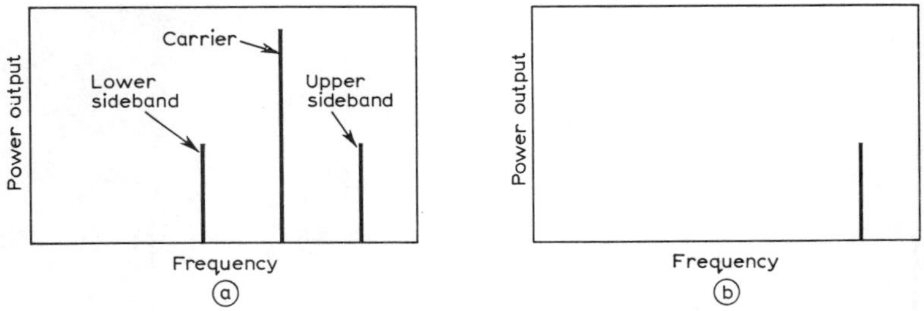

Fig. 4.13 (a) Spectrum of signal from a carrier modulated with a sine wave of one frequency, amplitude modulation used. (b) Lower sideband and carrier removed.

As far as a receiver is concerned, the carrier conveys no information and is completely redundant. If the receiver has a narrow bandwidth, only one *sideband* will be received, so that the other sideband is also redundant. Single sideband (SSB) transmitters send out one sideband with no carrier. Practically all the power used for transmission is used usefully in carrying information since none is wasted in supplying redundant signals.

The carrier is usually removed by modulating two carriers which are phase shifted by 90°, then mixing them so that the modulated sidebands add and the carriers subtract. The unwanted sideband may then be attenuated by filters.

At the receiver, the carrier must be inserted so that demodulation (detection) can take place; this is done by an oscillator within the receiver. A stable oscillator can be used, or the oscillator can be controlled by sampling the usable sideband and part of the attenuated sideband and producing an output proportional to the mean frequency.

CHAPTER FIVE

FREQUENCY LIMITATIONS IN UNTUNED AMPLIFIERS

FOR ANY TYPE OF AMPLIFIER, whether it is tuned or untuned (sometimes called aperiodic), there are definite upper and lower limits to the frequencies at which it can operate. In some cases modified circuitry enables us to extend the frequency limits of an amplifier; in other cases, such as occur at frequencies of several thousand MHz, specialised devices, such as travelling-wave tubes, must be used for amplification.

Chapter 4 dealt with tuned amplifiers and the methods used for obtaining wide or narrow bandwidths; this chapter deals with the problems of handling very low and very high frequencies in untuned amplifiers. This type of amplifier is necessary, for example, in oscilloscopes where waveforms of various frequencies or pulses of various widths and rise-times must be examined, and the amplifier must amplify all frequencies equally. The bandwidth required may be from d.c. (z.f.) to 25MHz, for some work it may have to be from 100Hz to 500MHz, and the circuit techniques used in making such amplifiers, called wideband untuned amplifiers, require some consideration.

Fig. 5.1 shows a typical *CR* coupled amplifier stage. It is fairly easy, once stray capacitances have been drawn in, to identify the reasons for the amplification falling off at low and at high frequencies. The coupling time constants into and out of the amplifier are differentiators, and so limit the response at low frequencies. The emitter time constant also limits gain at low frequencies because of the negative feedback caused by *R*3 when it is no longer effectively bypassed by the capacitor.

The stray capacitance, both of this stage and of the next one, across the load resistor, has an integrating action which cuts the response at high frequencies. The input also has a stray capacitance which has an integrating action. Bearing these limitations in mind, we shall now see what can be done to extend the frequency range.

Low Frequencies

Since it is the differentiating time constants which limit the gain of the amplifier at low frequencies, the logical first step to extending frequency response is to make these time constants as large as possible. This presents little difficulty as far as the time constants in emitter or cathode bypass circuits are concerned, or with valve G_2 decoupling, but becomes a problem when stages are coupled using very large capacitors or large resistors or both.

The size of grid resistor which can be used in a valve circuit is limited by the maker's specification because of the voltage drop caused by grid current. The input impedance of a transistor is governed by the base-to-emitter impedance, which is comparatively low, though bootstrap circuits can produce very high input impedances at the cost of greater complexity. A MOSFET can be used with extremely high values of input resistor (several thousand MΩ) and need have no bias.

In each case, the input impedance can be increased very considerably by the circuit device known as 'bootstrapping'. Bootstrapping involves feeding back signal in phase to the other end of an input resistor, so that the signal across the resistor is almost zero (mentioned briefly in Chapter 2).

If there is no signal voltage across a resistor, there is no signal current through it, and it behaves as if it had an infinitely high resistance. For amplifiers using valves or MOSFETs, the input resistance is limited by the resistor used to earth or bias the input terminal. The apparent size of this resistor can be increased by bootstrapping it, as for example in Fig. 5.2 where the input resistor is 10MΩ.

If the gain of the FET used as the input stage is 0·9, then only 0·1 of the input signal appears across the 10MΩ resistor, so that its apparent value is 10/0·1MΩ, or 100MΩ. If the gain of the first stage were 0·99, then only 0·01 of the signal is across the input resistor which would then seem to have an input resistance of 1,000MΩ. Note in the circuit shown that the drain of the FET has also been bootstrapped so as similarly to increase the impedance of the gate-to-drain capacitance.

In any bootstrap circuit, if the bootstrapping is carried out through a time constant, then the time constant used will determine the change of input impedance with frequency, so that bootstrapping the input of an amplifier will have no effect on the low frequency response unless the bootstrapping is through a d.c. coupling or with a very long time constant.

Transistor amplifiers require rather more elaborate bootstrapping,

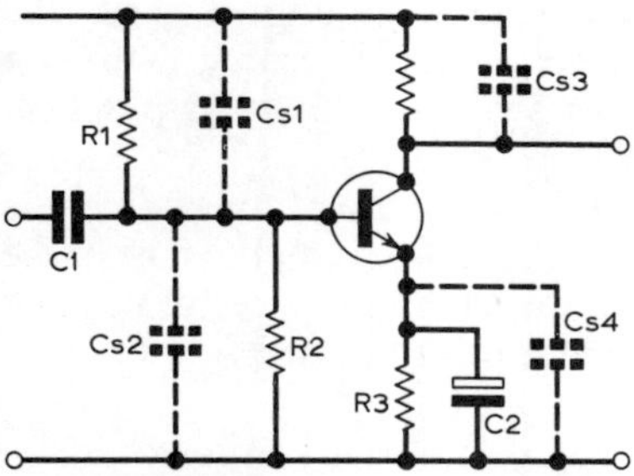

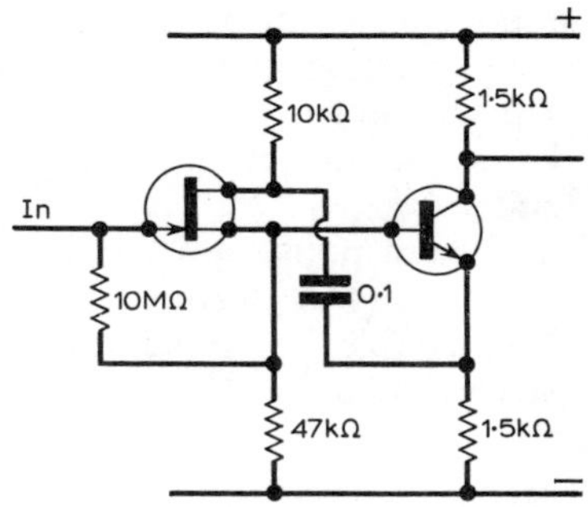

Fig. 5.1 (left) CR coupled stage. Cs indicates a stray capacitance.

Fig. 5.2 (right) Bootstrapped input stage using junction FET. Note the additional bootstrap connection from the transistor emitter to the FET drain, so bootstrapping the drain/capacitance at high frequencies.

since the limit to input resistance is caused by the base-emitter resistance rather than by the input resistor network used for bias. Using modern silicon transistors, quite high impedances can be achieved in common-collector or common-emitter circuits when low collector currents are flowing, but impedances of well above 1MΩ can be obtained only if both bias resistors and base-emitter junction are bootstrapped, as in the circuit of Fig. 5.2.

In audio amplifiers, thermal stability of output transistors is aided by the use of low value bias resistors in the base circuit. So as to preserve reasonable low-frequency response, it is common practice to bootstrap these resistors by a connection from the amplifier output. Some care has to be taken, when more elaborate bootstrapping arrangements are used, that the gain from the amplifier stages which are bootstrapped does not exceed the loss in the bootstrapping network, otherwise oscillation will take place, since the feedback is in phase. This is automatically safeguarded when common-collector circuits are used, since the maximum gain is one (unity).

As far as extending the size of coupling capacitors is concerned, the greater the capacitance the greater the leakage. A leakage of a few microamps in a valve circuit may upset bias conditions when it causes a voltage to appear across several MΩ of input resistance. The same leakage in a transistor circuit would have less effect unless the transistor were already operating with a very low bias current. A similar leakage in a MOSFET stage which caused too high a voltage to be placed on the gate would cause the destruction of the MOSFET.

There is a limit, therefore, to the size of coupling *CR* time constants which can be used even if unlimited space is available for large coupling capacitors. If transformer coupling is used, the primary inductance of the transformer is the governing factor on the low frequency response, but this again is subject to practical limits.

When response down to d.c. is required, the use of time constants for coupling or decoupling is not feasible; the amplifier must be direct-coupled, or some other technique such as d.c. restoration or the conversion of d.c. to a.c. (chopping) must be used.

Direct Coupling

When the signal coupling from one amplifier stage to another does not involve a time constant (apart from stray capacitance) of differentiation, the amplifier is said to be directly coupled. Some direct coupling circuits are shown in Fig. 5.3. In each case, the difficulty which must be overcome is of obtaining the correct bias

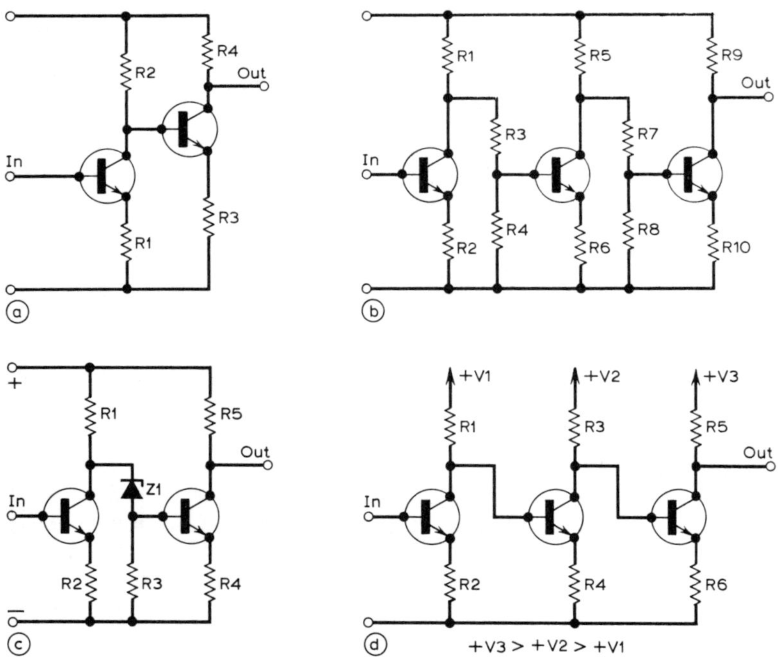

Fig. 5.3 Direct coupling: (a) R3 could be replaced by a Zener diode, (b) signal is also attenuated by the divider networks, (c) use of Zener diode in coupling, (d) use of separate stabilised power supplies.

conditions when coupling together electrodes which must be at least a few volts (d.c.) apart or, in the case of valves, even several hundred volts apart.

One method is to use a resistive potential divider from one output to the next input; this has the obvious disadvantage that the signal is stepped down in the same ratio as the d.c. Another method is to use a gas stabiliser (high voltages) or zener diode (low voltages) as the coupling element, since these act as very low impedances for signals but have a constant voltage difference between their terminals. Gas stabilisers have the disadvantage that they have a poor response to high frequency signals and may extinguish during large signals, causing non-linearity, but zener diodes are excellent and are widely used.

A third technique is to arrange the bias for each stage so that the stages can be directly coupled. This may require separate voltage supplies for each stage, with each supply stabilised, as any difference will be amplified as if it were a signal. Designers of direct coupled amplifiers have devised many ingenious methods of stabilising circuits against changes in the power supply voltages.

Direct Coupling in Op-amps

Operational amplifiers are always direct coupled, and the techniques used in these amplifiers are of interest. Fig. 5.4 shows a typical op-amp, the Motorola MC1530 with an open-loop (no feedback) gain of 72dB. Transistors Q_2 and Q_4 form a differential amplifier using as a common emitter load the transistor Q_3 which provides a constant current. The operating conditions of Q_3 are stabilised by the circuit (D_1, D_2, R_4) in its base.

The voltage gain of this stage is about 60, and this is the largest stage gain, ensuring that any voltage offset in the later stages can be amplified only by the smaller gains of these stages. Because of the separate positive and negative line voltages used, the inputs can be balanced about earth potential, and the outputs appear at a level of about $\pm 2{\cdot}2$V. The second differential stages, Q_4 and Q_5, have a voltage gain of about 20 and use a resistive emitter load.

The voltage level of signal at the collector of Q_5 is about 3·8V, and a level change is now needed to provide an output of about zero volts level. The level change is provided by the network composed of R_9, R_{10}, and Q_6, so that the emitter of Q_7 is provided with an output at about zero volts level, and drives the output pair Q_7/Q_8 with the output from the emitter of Q_7 and collector of Q_8. The gain of the output stage is approximately 5.

Chopper Amplifiers

Another approach to d.c. response is the 'chopper' amplifier. If the input of an amplifier is switched alternately between earth and a steady (or slowly varying) voltage, the input to the amplifier consists of square pulses at the frequency of operation of the switch and of peak-to-peak amplitude equal to the steady voltage to be amplified. The problem is now to amplify square pulses, much less of a problem than amplifying d.c.

The amplifier output consists of (more or less) square pulses which can be then reconverted to d.c. by rectifying or by switching alternately from one output to a paraphase output (synchronous operation). The switch, the key to the whole process, is known as a 'chopper' since it chops d.c. into square pulses, and it may be mechanical (a reed relay) or electronic (a valve or transistor).

The highest chopping rates are achieved with electronic devices, and transistors are much more satisfactory choppers than valves, since they perform much better as switches. Where very high input impedances are required MOSFETs may be used as choppers. High chopping rates have the advantage that large time constants are not required in the amplifier, and the size of the whole system can be kept down. Chopper amplifiers may be externally driven by pulses, or may be *self-oscillating*, where the amplifier itself causes the chopping to take place.

D.C. Restoration and Clamping

These are techniques for replacing a d.c. component into a signal after it has passed through a coupling which does not permit response down to d.c. The waveform must be one which has a definite d.c. level at some point in the waveform and the most obvious example is the TV line waveform which has as its defined level the tip of the synchronising pulse. If the defined level is one extreme of a waveform, the simpler technique of d.c. restoration can be used, which consists of connecting a diode between the signal path and a voltage level to which the waveform tip is to be restored.

Taking as an example (Fig. 5.5) a waveform containing a negative pulse whose tip is to be restored to -2V, the diode is connected in a fairly high impedance path fed by a capacitor. The polarity of the diode is such that signals of less than two volts level cause conduction. At any time when the diode is conducting, the current flowing will charge the capacitor C until the minimum level of

−2V is reached (assuming zero forward voltage in the diode).

Though the capacitor will now slowly discharge through the remainder of the circuit, other voltage pulses will again cause recharging. If the need is purely to replace a d.c. component in this way, the time constant of capacitor discharge is not critical and may be very large.

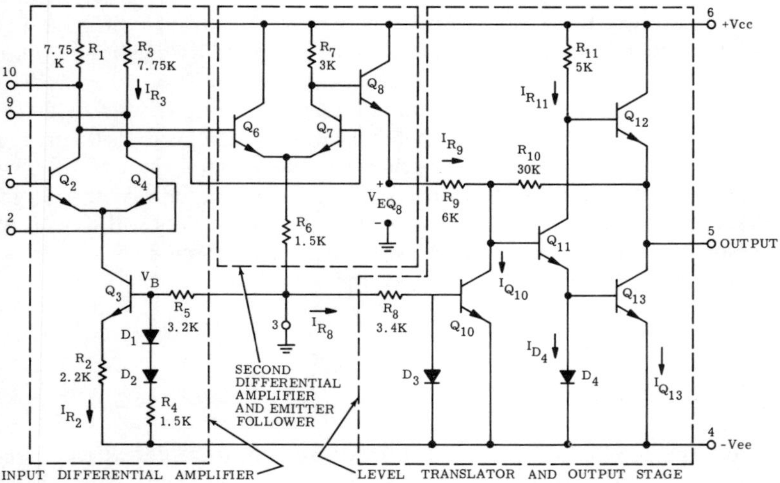

NOTES

(1) ALL DIODE FORWARD VOLTAGES AND BASE-EMITTER VOLTAGES ARE APPROXIMATELY 0.7 VOLTS.

(2) ALL BASE CURRENTS ARE NEGLECTED

(3) NOTATION: V_{BQ_3} IS READ AS THE POSITIVE TO NEGATIVE VOLTAGE FROM BASE TO GROUND FOR Q_3; V_{EQ_8} IS FROM THE EMITTER OF Q_8 TO GROUND, ETC.

Fig. 5.4 Circuit of Motorola MC1530 shown in portions indicated by dotted lines to separate the different stages.

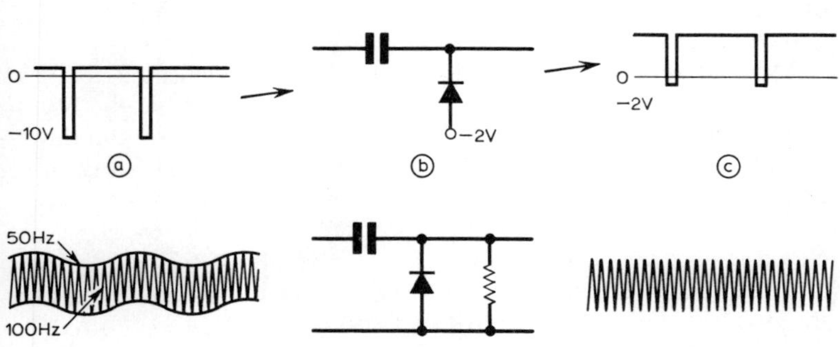

Fig. 5.5 D.C. restoration: (a) pulse waveform input, (b) circuit, (c) output restored to −2V level, (d) use for removing hum.

The technique is also useful for removing an unwanted signal which is at a lower frequency than the pulse repetition frequency. If, for example, a 1,000 p/S pulse waveform is affected by hum, the 50Hz signal can be removed by d.c. restoration, this time using a time constant which ensures that the diode conducts to a small extent on each pulse, so keeping the tip of each pulse tightly at the chosen potential and eliminating the hum.

Clamping can be carried out at any part of a waveform which is at a fixed time from a synchronising pulse or any other regular steep portion of waveform. Once again, the signal is connected to a selected voltage by a low impedance path, but in this case the connection is achieved by a switching circuit such as a diode bridge and is carried out by a pulse occuring at the selected time. Clamping can be bi-directional, meaning that the waveform can be shifted in voltage in either direction, and is a more precise method than d.c. restoration by a single diode, since it is not affected by anything which has happened to the tips of the waveform.

High Frequencies

The factors which limit the bandwidth of an amplifier at the *high frequency* end are the integrating time constants caused by stray capacitances, including those which exist within the valve or transistor, along with frequency limitations which take place within the valve or transistor and which are caused by the time taken for the current carriers to travel from one electrode to another. There is no way round the latter limitation except to use an active device capable to operating at high frequencies, but there are several ways of increasing bandwidth when the limitation is due only to stray capacitance.

Most of these methods involve sacrificing gain for bandwidth. For example, if the load of a stage is lowered in value, the bandwidth must increase because the time constant of the integrating circuit formed by the strays and the load is lowered. With any valve or transistor, there is a limit to this process, which can be calculated from knowledge of the amplification factor and the capacitances between electrodes. In many cases, the result of this calculation is tabulated in the valve or transistor data sheets as the 'Gain-Bandwidth' product, meaning that the gain of a stage using the device multiplied by its bandwidth can not normally exceed this amount.

The G-B factor is usually quoted in MHz; for example, a G-B

factor of 500MHz means that a bandwidth of 500MHz would be possible for a gain of 1; a bandwidth of 100MHz for a gain of 5; a bandwidth of 25MHz for a gain of 20, and so on. Remember that these figures apply to *one stage;* for two stages the gain is *squared* and the bandwidth halved; for three stages the gain is cubed and the bandwidth 1/3 of that for one stage.

In some cases it may be very inconvenient to use very low value load resistors, especially if large voltage swings are being used so that a very large bias current would be required. A negative feedback amplifier may be used, with anode loads of normal value to cope with the voltages being amplified, but with feedback causing the gain to be low. In such a case the bandwidth should be the same as it would have been if loads low enough to give the same gain had been used.

One vital difference, however, is that the feedback amplifier may show peaks of high gain in its response at high frequency and may even oscillate. This is due to the phase shifts in the time constants of the amplifier making the feedback veer towards positive feedback at high frequencies. For this reason, many amplifiers use a combination of fairly low loads and feedback with the feedback only applied over a few stages.

Slew Rate Limitations

Slew rate limitations may arise with feedback amplifiers. In any amplifier, the stray capacitances are constantly being charged and discharged by the signal voltages. For charging or discharging any capacitance, current must flow in or out of the capacitance, and this current must either be supplied by the active device or allowed to flow through the load resistor. If this current is limited to a value I, then $I = C \times$(rate of change of voltage), where C is the stray capacitance being charged or discharged. From this, then, the maximum rate of change of voltage is given by I/C (I in amps, C in farads) and is in volts per second.

Where valves are used this does not severely limit circuit design because even valves designed for high frequency operation are capable of passing fairly high currents. This is not necessarily true of transistors and is particularly limiting for integrated circuits where the problem of heat dissipation makes it difficult to use high values of current. In addition, the higher stray capacitances and the capacitive networks used in op-amp circuits make slew rate the main limitation in integrated wideband amplifiers.

Cathode and Emitter followers

The cathode follower (valve) or emitter follower (transistor) is a circuit of very large bandwidth, but with a voltage gain less than 1. The output is in phase with the input, and is taken from the cathode or emitter respectively and the input being at grid or base (Fig. 5.6). This circuit has a large number of useful features:

(1) The output is of very low impedance.

(2) Since the output is not taken from the anode/collector, which is earthed to signal voltage, there is no 'Miller Capacitance' across the input, which is therefore of high impedance.

(3) Because the output voltage *follows* the input voltage (hence the name), the stray capacitance between input and output (for example, the base-to-emitter capacitance) is not being charged and discharged to any great extent. For this reason, it behaves as if it had *very low stray capacitances.*

(4) Because of (3) it has very high bandwidth.

(5) Also, because of (3), it can be used as a stage between two voltage amplifiers. One amplifier drives the very low strays of the follower, and the follower with its low impedance easily drives the strays of the next amplifier.

(6) Though there is no *voltage* amplification, there may be very considerable *current* amplification, and emitter followers may be cascaded to provide very large amounts of current amplification.

The only serious disadvantage of the follower is that the response to a pulse which turns current off in the device is slow. Fig. 5.7(a) shows why; stray capacitance across the load charges up quickly due to the current flowing in the transistor, but when the tail-end of the pulse arrives, the stray capacitance is unable to discharge through the load resistor at the rate at which the pulse voltage is dropping. The transistor cuts off, and the stray capacitance dis-

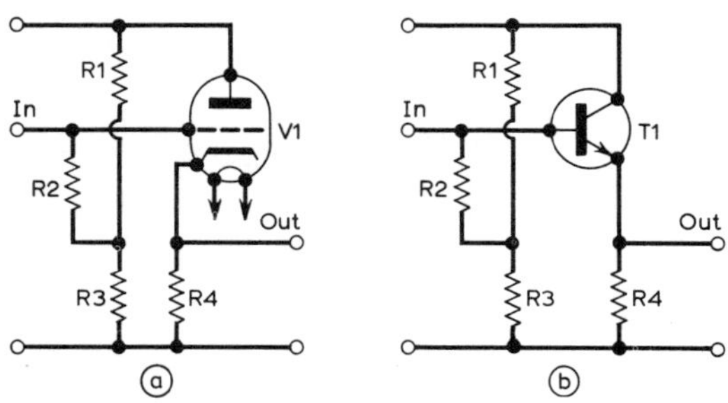

Fig. 5.6 (a) cathode follower, (b) emitter follower.

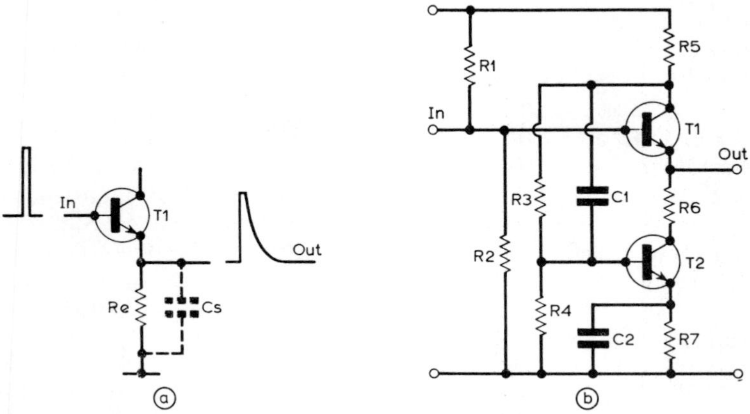

Fig. 5.7 (a) trailing response at negative side of pulse caused by discharge of stray capacitance Cs through Re when T1 is cut off. (b) compound follower uses second transistor to discharge strays.

charges slowly to earth. This can be overcome by using compound cathode or emitter followers of the type shown in Fig. 5.7(b).

The output impedance of a cathode follower is approximately $1/Gm$. If Gm is in mA per V, $1/Gm$ is in kΩ so that a valve of $Gm = 5$ would have an output impedance of 200Ω as a cathode follower. For a transistor, the output impedance is very low, about 25Ω for most transistor types.

The output resistance of the emitter follower is given by $1/Ro \simeq 1/Re + hfe/(Rs + Rbb' + Rb'e)$ where Re is the emitter resistor in ohms, hfe is the current gain, Rs is the source resistance, and Rbb', $Rb'e$ are internal transistor resistances. This expression cannot be so easily simplified as the corresponding expression for the cathode follower, but the factor $1/Re$ is generally small compared to $hfe/(Rs + Rbb' + Rb'e)$ and, if the source resistance is high compared to the transistor input resistance, this simplifies again to $Ro = Rs/hfe$.

Two points are worth noting. Firstly that a collector load resistor makes no difference to the output resistance at the emitter, and the emitter resistor itself is of little importance if its value is several hundred ohms or more, and secondly that the *source* resistance has such an important effect.

Followers with Gain

If an operational amplifier is wired so that the + (or inphase) input is used, this is often referred to as a 'follower with gain'.

Using the circuit of Fig. 5.8(a), the gain is given by $(1 + R2/R1)$. The input and output resistances depend on the loop gain, which is $R1/(R1+R2) \times A_0$, where A_0 is the open-loop voltage gain, usually 5,000 or more.

For example, if the output resistance of an op-amp is 2kΩ, the open-loop gain is 5,000 and $R1 = 10$kΩ, $R2 = 40$kΩ, then the loop gain becomes $10/50 \times 5{,}000$, which is 1,000, and the output resistance becomes R_0/(loop gain), which is 2kΩ/1,000, equal to 2Ω. Similarly, the input resistance of the follower with gain becomes $R' \times$ loop gain, where R' is the differential input resistance.

The op-amp can also be connected with the inverting input connected directly to the output and the input taken directly to the non-inverting terminal. Again, the input impedance is very high and the output impedance very low, but some care has to be taken with input signals, as they are applied in almost equal amplitude to both input terminals, one directly and the other through the feedback loop. Because of this, the common mode input voltage of the op-amp must be adequate to handle the input voltages used. The follower-with-gain is not so restricted, as only a fraction of the output is fed back.

Inductive Compensation

This is a method of increasing the gain-bandwidth product for an amplifier stage, and is illustrated in Fig. 5.9. By inserting an inductor in series with the amplifier load resistor, the stray capacitance can be 'tuned out'. The effect of the inductor along with the stray capacitance is to form a parallel tuned circuit which has maximum impedance at the frequency of resonance.

If the resonant frequency is made to be slightly higher than the frequency at which amplification would normally be 3dB down, then a level response can be maintained to a wider bandwidth. At first sight it might appear that frequency response could be greatly increased by this method, but the load resistor determines the Q of the final arrangement so that the increase in bandwidth is limited. More complex arrangements are possible; another inductor may be used to couple one stage to the next so that the input capacitance of the next stage is tuned out separately, but in general the cost of using such circuits soon exceeds the benefits.

Inductive compensation is most valuable where the number of stages which can be used is limited and where feedback is not in use. With several stages, the combined effect of the inductors is to cause 'ringing' (oscillating overshoots) on squarewaves and an uneven frequency response. Feedback cannot be readily used with an inductively compensated amplifier because of the violent changes in phase which take place near resonance.

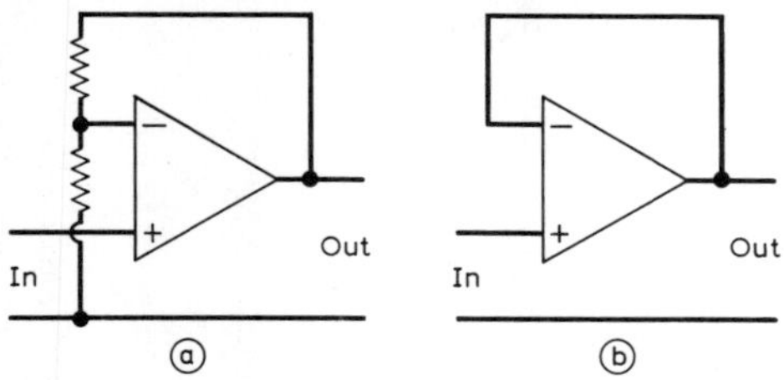

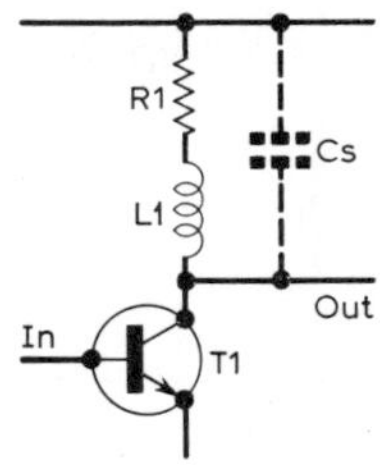

Fig. 5.8 (left) Op-amp followers: (a) follower-with-gain, (b) 100% feedback follower.

Fig. 5.9 (right) Inductive compensation.

Emitter/Cathode Compensation

Where an even frequency response is of more importance than very high gain, as is often the case in wideband amplifiers, the transistors or valves may have their emitter/cathode resistors un-bypassed, so introducing some negative feedback and reducing gain. If now a small value of bypass capacitor is added to each resistor in parallel, the gain must rise at frequencies where the impedance of the capacitor is comparable with the emitter/cathode resistor value.

If the time constant of the *CR* combination in the emitter/cathode circuit is made equal to that which causes the loss of gain at high frequencies (the load/stray capacitance time constant), then the two will counteract each other until the loss due to strays becomes more than can be compensated by the increase in gain. There is no increase in Gain-Bandwidth product by using this method, but it is reasonably trouble-free and easy to set up.

Examples of Wideband Amplifiers

The circuit of Fig. 5.10 is a portion of the video amplifier for a high-quality oscilloscope (Advance Instruments). The bandwidth is from d.c. to 5MHz (3dB down) and the gain of the whole amplifier is such that the maximum sensivity of the oscilloscope is 100mV/cm. The portion shown is of the latter half of the amplifier, omitting the input attenuator, cathode follower, emitter follower and a gating network which shifts the position of the trace and applies signals from another preamplifier so that two traces may be shown on a single-gun tube.

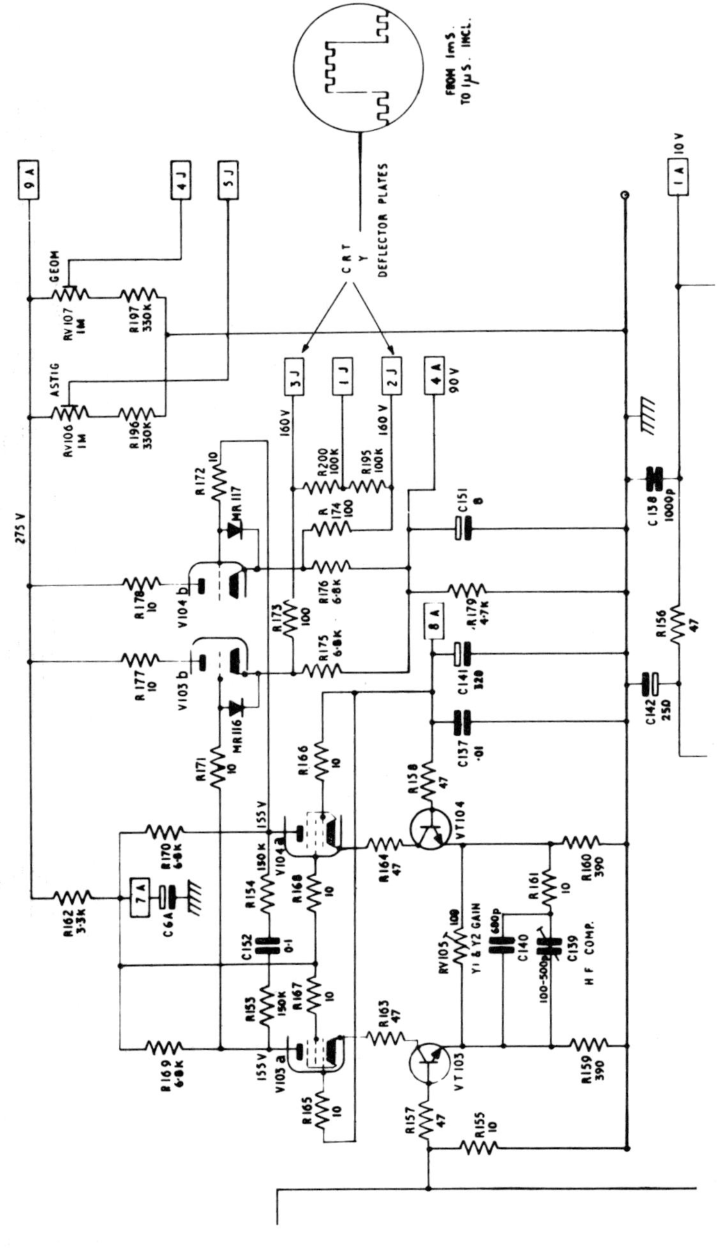

Fig. 5.10 Portion of Y-amplifier of type OS25 Oscilloscope. (courtesy of Advance Electronics Ltd.)

The amplifier is an interesting example of mixed transistor and valve circuitry with a long-tailed pair of transistors driving a pair of pentodes in cascade. *VR*105 sets the gain of the amplifier by regulating the amount of signal to the long-tailed pair. *C*140 and *C*139 increase the gain of *VT*103/*VT*104 at high frequencies by by-passing *VR*105, so achieving high frequency compensation by feed-back as previously described. Note the thorough decoupling of the base of *VT*104 which serves also as the bias source for the grids of the pentodes *V*103a and *V*104a.

These pentodes use the low value load resistors (6·8kΩ) *R*169/*R*170, and feed cathode followers *V*103b/*V*104b directly. The net-work *R*153, *R*154, *C*152 between the anodes of *V*103a/*V*104a is a balancing network which helps to counteract, by feedback, the differences in the amplification of the two valves caused by inevitable differences in characteristics. The cathode followers have restoring diodes *MR*116, *MR*117 at their grids and 6·8kΩ loads in the cathodes. The lower ends of the loads are taken to a common 4·7kΩ resistor, so that excessive bias currents do not have to pass through the 6·8kΩ loads.

Fig. 5.11 shows a quite different amplifier, a pulse video amplifier of a marine radar set (Decca Radar Ltd.) *Tr*102 acts as amplifier and mixer, inserting various marker signals into the received video signal. Emitter compensation is used in this stage, along with a 1·5kΩ collector load. *Tr*108 acts as an emitter follower, amplifying current to drive the base of *Tr*109, which uses emitter compensating capacitor *C*102, low value collector resistor *R*154 and inductor *L*101 which compensates for the capacitance of the cathode of the

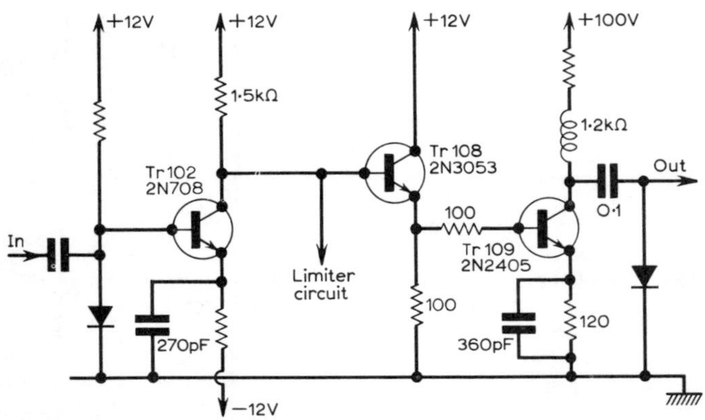

Fig. 5.11 Simplified video amplifier of Decca "Transar" marine radar, omitting mixing and limiting circuits. Note the simplified bias arrangements possible when pulses of only one polarity are handled.

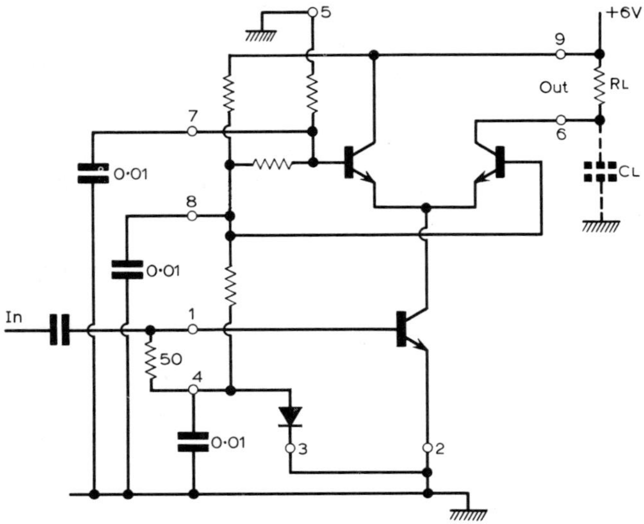

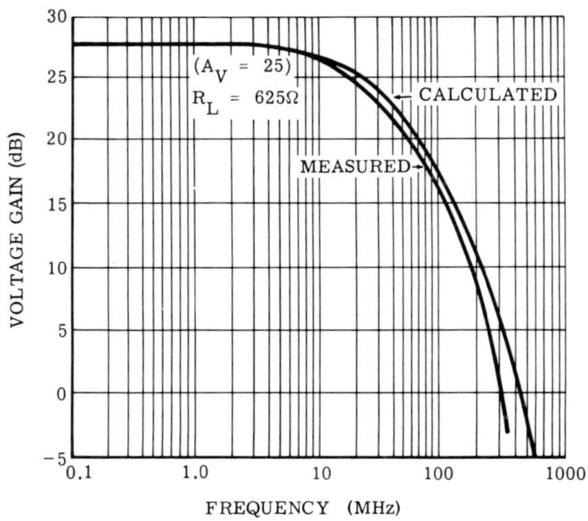

Fig. 5.12 IC wideband amplifier type MC1550, and response curve, assuming load of 625Ω and 5pF. (*Motorola Ltd.*)

c.r.t. to earth. The gain of this amplifier is 40, and the rise time to the 90% level is 0·05μS, corresponding roughly to a bandwidth of 7·5MHz.

By contrast, Fig. 5.12 shows the use of a Motorola MC1550 IC amplifier as a video amplifier with a gain of about 28dB over a bandwidth in excess of 20MHz. The input is applied between base and emitter of *Q*1 and the output taken from the collector of *Q*3, so that *Q*1/*Q*3 act as a common-emitter/common-base pair having very low feedback compared to a single transistor of the same gain. Operating conditions are set by the voltage applied to the base of *Q*2.

Magnetic Amplifiers

Magnetic amplifiers depend for their action on the variation of the permeability of a magnetic core as a result of variation in the current passing in a coil wound round the core. The simplest possible magnetic amplifier circuit is shown in Fig. 5.13. The circuit consists of two windings on a metal core. One winding, (3–4) is connected in series with the load to an a.c. supply. The other winding, the control winding (1–2), is connected through an inductor to a d.c. supply.

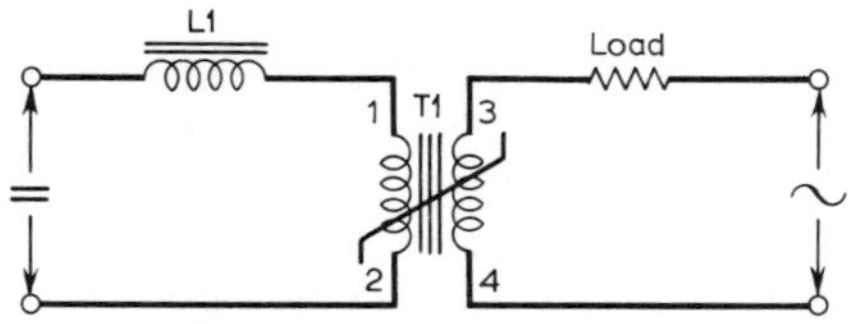

Fig. 5.13 Principle of magnetic amplifier.

When there is no d.c. applied, the winding 3–4 behaves as a high value inductance, and greatly limits the current which can flow in the load. When d.c. is applied to the control winding, the magnetism of the core increases, and can be increased to such an extent that the core eventually saturates, meaning that no further change in current will cause any change in magnetism. However the inductance of a coil depends on the change of magnetism in the core, so that, at saturation, there is no longer any appreciable inductance.

A comparable case is by increasing the bias current in the base of a transistor until the voltage at the collector is almost equal to the emitter voltage, and no further signal or bias can change it. In the case of the magnetic core, there is a region before saturation is reached where small changes in the control current cause large changes in the inductance of the main winding. In this way, a current in the main winding can be controlled by a current in the control winding. The main current is a.c., and the control current is d.c. or very slowly varying a.c.

The amount of gain in terms of a.c. change to d.c. change can be very high (100 to 1,000 or more) in one amplifier, and several stages can be cascaded by passing the a.c. from one stage into a rectifier and then using it as the control current for another stage. Magnetic amplifiers are extremely useful for high gain amplification at very low frequencies (less than 1/10 of the frequency used in the load winding) and are used in the control of high power circuits, since the dissipation in the magnetic amplifier can be quite small.

Varactor Parametric Amplifiers

The magnetic amplifier is an example of a method of amplification which uses a control signal to vary the value (parameter) of a component in the load circuit. In the magnetic amplifier, the parameter being varied is inductance. The name 'parametric amplifier' is usually reserved, however, for a method of amplification which is extremely useful at frequencies ranging from a few hundred MHz to a few thousand MHz and which uses, as the variable parameter, the capacitance in a tuned circuit.

A semiconductor diode, reverse-biased, has a capacitance which varies with the reverse voltage applied to it. A diode made to emphasise this feature is known as a varactor (*var*iable re*actor*) diode (or *varicap diode*), and the variations of capacitance can follow a signal imposed on the reverse voltage. If such a diode is made part of an oscillating circuit, and such a signal is applied, then, for a low frequency input a frequency modulated output is obtained.

For higher frequency signals, there is an output at a frequency equal to the difference between oscillator and signal frequencies (as is also the case with any mixer) but of amplitude considerably greater than the signal frequency. There is also an output at the sum of the frequencies which is generally tuned out and unused.

Travelling Wave Amplification

This is an attempt to remove the gain-bandwidth restriction. Normally, if two amplifying devices are connected in parallel there is no increase in G-B product because the capacitance has doubled as well as the gain. If the inputs of the amplifying stages are connected along a delay line of suitable impedance, the signal reaches each input in turn, the time delay being determined by the construction of the line, which can include stray capacitances.

The time delay means that the input capacitance of one device does not add to that of another. The output signals will also occur in sequence, and if the outputs are also connected by a delay line of similar characteristics, the signal from the first output will reach the end of the line just as the input signal reaches the last amplifier stage, and the outputs will add.

Using this method, almost any desired G-B product can be obtained if enough amplifying devices are used and sufficient sections of delay line wound. Bandwidths of over 1,000MHz at gains sufficient for oscilloscope use have been successfully obtained.

CHAPTER SIX

OSCILLATORS

AN OSCILLATOR IS ANY DEVICE OR CIRCUIT which produces an output signal waveform so long as d.c. is supplied to it. In some cases the oscillator may be started and stopped by another waveform, but the basic job of converting d.c. to a signal waveform is the same. Oscillator circuits consist of three types of circuit element already familiar: an amplifier, a positive feedback loop, and a time constant.

Oscillation takes place when feedback is in phase, and when the loop gain is greater than one, so that the amount of signal fed back is sufficient to maintain or increase the output signal. Oscillating active devices contain all these elements of amplification, feedback and time constant in one package; for example, the klystron or the backward wave oscillator.

Oscillators in common use are of two types—*sinusoidal oscillators*, which generate waveforms of approximately *sinewave* form, and non-sinusoidal or *relaxation oscillators* which generate all other waveshapes of a regular nature, such as squarewaves and triangular waves. The name, relaxation, comes from the sequence of events in a squarewave oscillator when the voltage rises, 'relaxes' back to a rest voltage, drops suddenly, and 'relaxes' again.

In linear amplifiers, the response to sinewaves is an important measurement but in the pulse and other non-linear amplifiers used in industrial electronics, other waveforms are more important. For this reason, relaxation oscillators are of greater practical importance, and we consider them first.

Relaxation Oscillators

The main circuit difference between sinusoidal oscillators and relaxation oscillators is that much larger amounts of positive feedback are used in the latter. The same circuit may give *either* type of waveform, depending on how the amount of feedback is

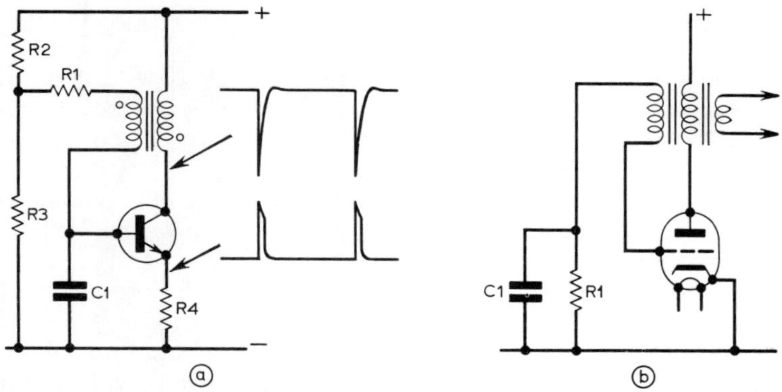

Fig. 6.1 Blocking Oscillator: (a) Transistor version. If R1 is too small or if the voltage at the junction of R2 and R3 is too high, older types of transistors may not oscillate. R4 should be small, about 100Ω. Time between pulses is about 2/3 R.C. seconds (R in MΩ, C in μF); (b) Valve circuit, showing use of extra winding on transformer for pulse take-off. R1 should be large to avoid excessive grid current.

controlled. The large amount of feedback used in a relaxation oscillator causes voltages and currents to rise and fall very rapidly when valves or transistors are amplifying, but for most of the time the active devices are *not conducting* or are *bottomed* (conducting so hard that no extra signal at the input can cause more current to flow). For example, if the collector of a transistor in a relaxation oscillator is drawing current, the positive feedback signal causes the base to draw more current, which in turn makes the collector draw more current and so on.

Positive feedback makes any change from full current to no-current or back very rapid, but can have no effect when no current is being drawn (no current, no signal) or when full current is drawn (bottomed state, no signal); the change-over in the no-current or bottomed condition is then done by the time constants. Note that the 'instability' of relaxation oscillators enables them to be synchronised (kept in step) with other sources of waveforms.

The *blocking oscillator* is a relaxation oscillator using a single amplifying stage, a transformer and a *CR* time constant. For positive feedback, the signal fed back to the input of the amplifying stage must be in phase with the signal which was there already, but a single stage amplifier (valve or transistor) inverts the signal fed to it, and some device must be used to re-invert the signal at the collector before feeding it back to the base. In the blocking oscillator, this device is the transformer, and the circuit is shown in Fig. 6.1, with the waveforms at various points in the circuit.

Blocking Oscillator Operation

On switching the circuit on, the voltage at the base causes an increase in current on the collector and a voltage is developed across the primary of the transformer. The transformer is connected so that the voltage which is induced across the secondary increases base bias still higher, causing current in the collector circuit to increase even more until bottoming occurs, and no further change in current can take place.

When this happens, the voltages across both primary and secondary of the transformer collapse (as only *changes of current* cause a voltage to appear), the base current decreases, causing the collector current to decrease and so putting a reverse voltage on the transformer which results in a rapid shut-off. Because of the amplification, the voltage cutting off the base is much greater than is needed (even when a step-down transformer is used), and the

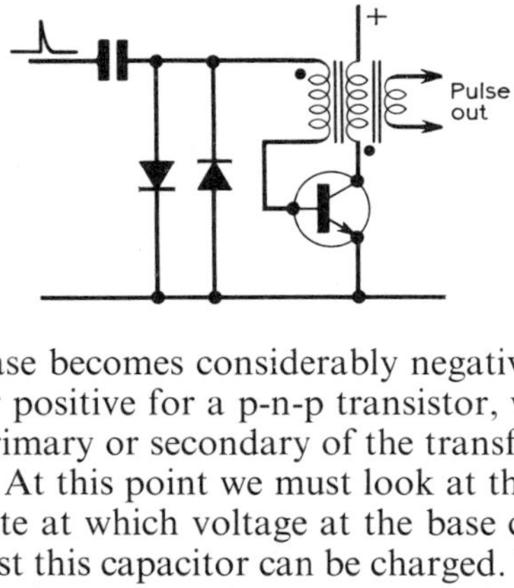

Fig. 6.2 Triggered blocking oscillator—the diodes prevent more than one cycle of oscillation for each trigger pulse.

base becomes considerably negative for a valve or n-p-n transistor or positive for a p-n-p transistor, with no current flowing in either primary or secondary of the transformer.

At this point we must look at the effect of the capacitor $C1$. The rate at which voltage at the base can be changed depends on how fast this capacitor can be charged. When the current was increasing or decreasing, the current to charge or discharge this capacitor came from the active device, either the grid-cathode of the valve, with grid positive, or the base-emitter of the transistor, in conduction. Either way, the impedance was low.

With the base biased off, however, the capacitor must charge mainly through the resistor $R1$, and it will charge at a rate determined by the time constant $C1R1$ until some current begins to flow in the collector circuit. This happens when a valve grid reaches cutoff voltage or when a transistor base starts to draw forward current, when the whole cycle repeats itself.

The output of the blocking oscillator consists of narrow pulses ($1\mu S$ wide or less) with a comparatively long time between, limited only by the longest time constant which can be used. The time constant can be longer in a valve or MOSFET circuit, because there is less leakage between grid and cathode when a valve is cut off than there is between base and emitter of a cut-off transistor unless a bootstrap circuit is used. MOSFETs are capable of very long time constants.

At the other end of the pulse, if the time constant is very long, the pulse portion cannot be very short, since the capacitor must change its voltage while the valve/transistor is drawing current, and the current which must flow to the capacitor has to flow through the base-emitter path, which can supply only a limited current.

In practice, the narrow pulses, which recur at a fixed time, can be used as timing pulses for timebases, for counting purposes, or as radar signal generators. The circuit can be constructed so that the pulse width depends mainly on the supply voltage (and the construction of the transformer) and this style of blocking oscillator, shown in Fig. 6.2, is well suited to triggered use, where each trigger pulse at the base causes an output pulse. Note that it is particularly easy to obtain an output, or several outputs in different phases, by using additional windings on the transformer.

The interval between the pulses when the voltage on the capacitor rises slowly can be used as a timebase in itself, for the voltage is approximately a sawtooth. A blocking oscillator is frequently used in television receivers to generate a timebase which is fed to an amplifier to supply the scan coils; in this case the curvature of the waveform is an advantage because opposite curvature occurs in the amplifier feeding the scan coils.

Complete failure of a blocking oscillator is unusual and must be due to failure of valve or transistor, of transformer (very seldom), or of the resistors and capacitors. More usual is a change of time constant due to change in value of the resistors.

Oscillators With Two Active Components

The Blocking Oscillator and the Transitron are the only commonly used relaxation oscillators with one active device. A third type is possible with transistors using the common-base connection and feeding back from collector to emitter, but this is used mainly for sinusoidal oscillations with an *LC* time constant.

The main difference between most relaxation oscillators and sinewave oscillators is the use of an *LC* time constant (with its high *Q*) for sinewave oscillators. The high *Q* network forces the oscillator to work at one frequency, instead of the mixture of frequencies which makes a squarewave square. This is a good distinguishing point, but it should be remembered that there are *CR* networks which have a comparatively high *Q*. (Twin T etc.).

The Multivibrator

There are several ways of arranging two transistors or valves so that relaxation oscillations are produced, and every circuit is some type of *multivibrator*. We shall outline the working principles briefly, relying on knowledge gained from the oscillators seen so far to understand the processes of rapid change due to positive feedback (regeneration) and of charging and discharging capacitors.

Although the treatment will be less thorough than with the previous cases, the working principles of the multivibrator family must be *thoroughly* understood if any headway is to be made with pulse circuits in Chapter 7. Students are recommended to make up a multivibrator with very large time constants so that voltages can be measured at various portions of the circuit.

Cross-coupled Multivibrator

Fig. 6.3 shows an example of this type of circuit, which consists of a two-stage *CR* coupled amplifier (whose output is in phase with its input) with the output *CR* coupled back to the input. During the period of regeneration, one device is being shut off and the other turned on, and the time between this and the next changeover is

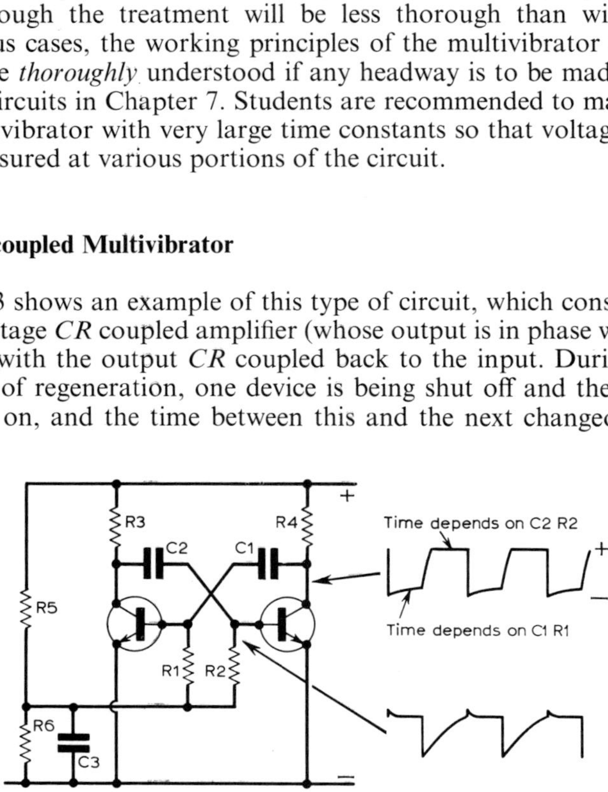

Fig. 6.3 Cross-coupled MV or Astable Flip-flop, with waveforms at one transistor. Biasing off one transistor produces a triggered MV or flip-flop, giving one squarewave for each trigger pulse. The time constants C1R1, C2R2 need not be equal, they each control one half of the squarewave. The circuit may not oscillate reliably if the base current in the ON condition is too high.

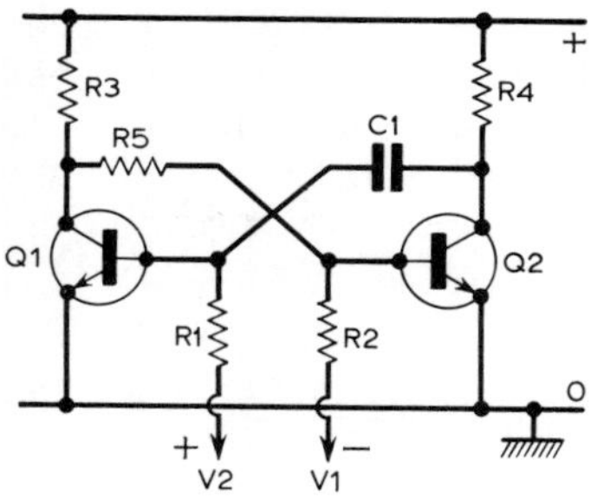

Fig. 6.4 MV with one direct coupling. Depending on the bias voltages, V1 and V2, this may be continuous (free-running) or triggered. For triggered operation, V2 should be zero and V1 negative enough to allow Q2 just to conduct. A positive trigger on the base of Q2 will then give an output of one square wave whose time is decided by the time constant C1R1.

determined by the time constant at the base of the transistor which is turned off, while the base of the transistor which is turned on is held just above emitter voltage by diode action, the voltage difference being the normal forward voltage across the diode.

Note the difference in the base waveform for the two transitions. The negative portion is a differentiated squarewave, whereas the positive portion has been clipped by diode action. If the two coupling time constants are equal, the output at each collector is a symmetrical squarewave—its negative portion takes the same time as its positive portion.

Because each coupling capacitor is charging from a negative voltage towards earth, the rate at which voltage rises may be slow when the device is turned on because of the shape of the capacitor-resistor charging curve. This causes the time between changeover to be uncertain; it varies from one change over so that an oscilloscope trace of the output of this oscillator shows the vertical portions of the squarewave *jittering* from side to side. This jitter can be greatly reduced if each grid/base is returned to a higher voltage. The capacitor is then charging more rapidly at the point when the other stage turns on and small changes in turn-on voltage do not cause large changes in turn-on time.

The cross-coupled MV can be used with a wide variety of time constants, though very long time constants are difficult with transistors, because of leakage, and very short time constants cause trouble because of stray capacitances and the high currents which can flow in grid/base circuits if the resistors are of low value. Fig. 6.4 shows a variety of cross-coupled multivibrators in which one coupling uses no time constant, but is a direct coupling. In this case the circuit must be adjusted to keep bias conditions correct.

In all cross-coupled MVs, the output from one anode/collector is exactly the inverse of the output from the other, so that this type of circuit is extremely useful when simultaneous switching pulses are required, one to be switching on, one off.

Emitter-coupled Multivibrators

This type of circuit (Fig. 6.5) uses one d.c. coupling between emitters and one *CR* coupling from collector to base. Remembering that the emitter always follows the base voltage, the stages in one cycle are as follows: if base(1) is going positive, emitter(1) is also going positive, cutting off *Tr*2 and causing its collector voltage to rise fully positive. The base(1) voltage then decreases as *C*1 discharges, the emitter voltage follows until *Tr*2 turns on. When this happens, collector(2) decreases in voltage, forcing the voltage at base(1) down and cutting off *Tr*1. The current in the emitter resistor is now taken entirely by *Tr*2 until *C*1 charges up again to the point when base(1) can take current again.

This circuit gives a very useful waveform at the emitter, consisting of a pulse at every changeover. As this negative pulse has a very sharp leading edge it can be used as a timing waveform (for example, to start one circuit operating just before the squarewave from the collector reaches full volts). The collector waveforms are fairly good squarewaves. The grid/base resistor in this circuit cannot be

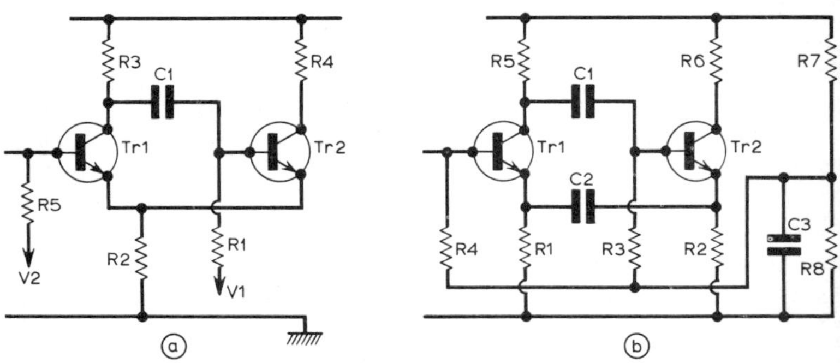

Fig. 6.5 Emitter coupled MVs. (a) When V1 = V2, about 2v, this free-runs giving a squarewave. If either V1 or V2 is earthed and the other is positive, the circuit becomes a triggered MV or flip-flop. Note that the base of Tr1 and the collector of Tr2 are left free for external connections; (b) Using two time constants. Again, one base and one collector are unused and may be externally connected.

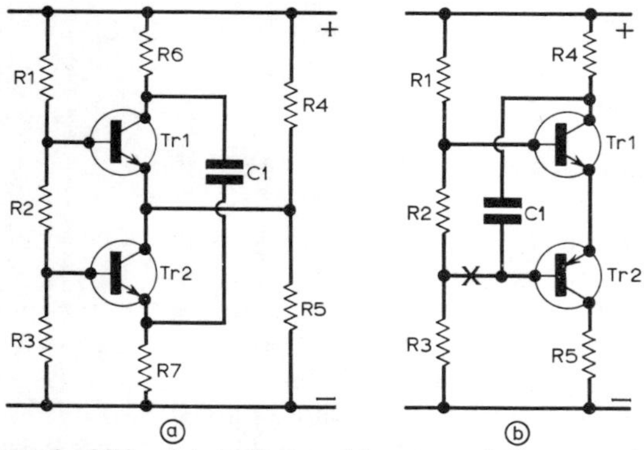

Fig. 6.6 Cascode MVs: (a) Both transistors are of the same polarity. The wavetimes depend on C1, R6 and R7; (b) Complementary symmetry oscillator. The voltage drop across R2 must equal the drop across both base-emitter junctions in series. Greater time constants can be obtained by inserting additional resistance at point X.

returned to a high voltage if continuous oscillations are required as this would cause *Tr*1 to switch on permanently unless switched off by a pulse applied from another circuit. This principle will be described, however, in the next chapter.

Fig. 6.5(b) shows a type of emitter coupled MV using two time constants, one of them in the emitter circuit. This type is usually made with the emitter time constants short (and equal) and the base time constant long, so that the circuit is entirely controlled by the emitter time constant of *C*2 and *R*1 or *R*2, whichever is in the emitter of the transistor which is shut off.

Cascode-coupled Multivibrators

Fig. 6.6(a) illustrates this type of MV using one d.c. coupling and one time constant. Once again, either one transistor or the other is conducting and holding the other one non-conducting, either through the time constant connection or through the emitter-collector connection. The resistor network *R*4 and *R*5 forms a current path through which each transistor can conduct when the other is shut-off, and a squarewave is obtainable (at low impedance) from the junction of *R*4 and *R*5 (compare the power amplifier circuit of Chapter 3).

Fig. 6.6(b) shows a cascode circuit using one p-n-p and one n-p-n transistor. This circuit is unusual because *both* transistors turn on and shut off together; the action is similar to a valve circuit (the transitron) which is now obsolete.

Multivibrator Design Variations

Although, basically, MVs of the types already described are widely used, many variations on the design are used to improve the performance. *Catching diodes* may be used in many possible positions to prevent voltages exceeding some preset limits, or to avoid having large current flowing through the active devices used. Some examples of the use of catching diodes are shown in Fig. 6.7, where the output waveform is clipped, and the timing waveforms are caught at voltages determined by the voltages to which the diodes are returned.

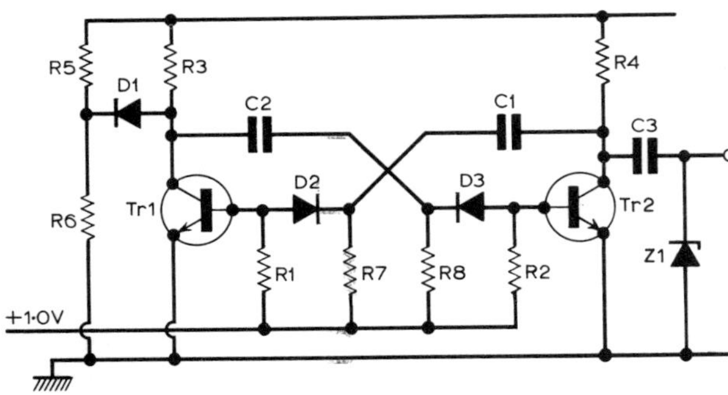

Fig. 6.7 Use of catching diodes. D1 limits the amplitude of the squarewave used to switch Tr2. D2 and D3 limit the loading of the collector circuit by base current, and Z1 clips the output wave and ensures a flat top of controlled voltage. The forward conduction of Z1 also gives d.c. restoration. R7 and R8 are large value resistors which provide discharge paths for C1 and C2.

At high speeds of switchover, it becomes more and more difficult to drive the coupling capacitor(s) of a multivibrator. The current required to change the voltage across a capacitor depends on the size of the capacitor and the rate of voltage change: (Volts change)/Time. When very rapid switchover is required, the current can usually be supplied by the collector of a transistor which is switching *on*, since the current is being supplied by the active device.

When a transistor is switched off, the rise of voltage is caused by the coupling capacitor charging through the collector load resistor, and the rate of rise of voltage depends on the size of the capacitor and the load resistor. There is a limit to how small a load resistor can be made if a sufficiently large voltage step is to be obtained from it at each switchover without passing excessive current.

One solution is to use an extra stage for charging and discharging the coupling capacitor. The usual method is shown in Fig. 6.8; each MV stage drives a d.c.-coupled emitter follower, and the current of the emitter follower drives the coupling capacitor. This problem, incidentally, is closely related to that of slew rate in amplifiers.

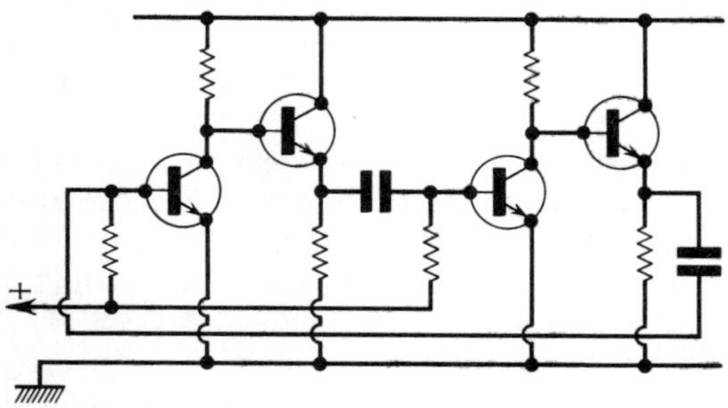

Fig. 6.8 Obtaining faster rise time by using emitter followers between the MV stages.

Faults in Multivibrators

Any o/c failure in coupling or o/c faults in active devices results in both collectors being permanently at line voltage. Any s/c failure in active devices or in coupling capacitors causes one (at least) active device to be permanently bottomed. If an oscilloscope test shows no output, a voltage test on collectors on the basis described should indicate the nature of the fault.

When reading such voltages, it is good practice to use a high impedance meter. If the MV is one which runs at a high repetition rate (1MHz or more), it is often necessary to have a 10kΩ resistor as a probe load on the meter or oscilloscope to avoid loading the circuit with the capacitance of the meter. Note that a 10kΩ resistor used in series with a Model VIII Avo causes a reading error of 0·5V *low*, which can have the effect of making a multivibrator run with unequally spaced positive and negative portions.

Sinusoidal Oscillators

Sinewave oscillators may be constructed using either *LC* or *CR* time constants. Normally, *CR* time constants are used only for lower frequency oscillators, for which inductors would be large and clumsy, especially compared with transistors, and require special circuitry to prevent them generating squarewaves due to the low Q of the CR time constants.

For frequencies above 100kHz, oscillators mainly use LC time constants, and two oscillator types are recognised: two-terminal oscillators, where only two connections are made between the active device and the time constant and four-terminal oscillators using four connections. The difference will become apparent as we examine various oscillators, although no attempt will be made to list every possible type of oscillator, since the working principles do not greatly differ.

As stated earlier, oscillators must use positive feedback, with a loop gain greater than 1, and there must be no phase shift between output and input. In practice, loop gains between 1·5 and 4 are used in oscillator designs, and the amplitude of the oscillation in the first few cycles increases until the loop gain decreases to a value of 1. This reduction of loop gain can be caused by bottoming, which has the same effect as reducing the load resistor, although this method cannot always be used, as for example, in self-oscillating mixers where bottoming would cause severe distortion of the incoming signal also.

If oscillation is allowed to cause the operating bias to shift more negative, conduction will take place only over part of the signal, so reducing the gain again to a value of 1. This second method of control is called *conduction angle control*. The circuit differences are small: the bottoming type is generally designed with the emitter connected directly to ground and the bias applied through a resistor to line voltage, and the conduction angle type uses a bypassed resistor in the emitter and a voltage divider supplying the base bias.

If the loop gain or decoupling capacitor values are too high, a blocking action called 'squegging' may occur. This causes the oscillator to operate for a few cycles, cut off, operate again, as if the sinecurve were modulated by a squarewave. Where this occurs, decreasing either loop gain or decoupling or both is effective as a cure; an undecoupled resistor in series with the emitter is also effective.

Oscillator circuits will change frequency (i.e., drift) slowly, and it is often desirable to reduce this change to a minimum. Where the change is caused by changes of temperature, the cause is generally in the transistor or the inductor. The use of undecoupled emitter resistors reduces the drift caused by the transistor (because mainly of variation of base resistance), and a capacitor with negative

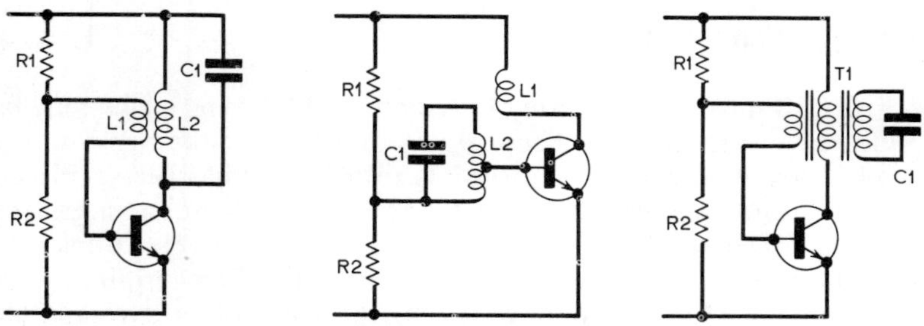

Fig. 6.9 Two-winding oscillators.

temperature coefficients may be used to counteract the effect of change of inductance.

Lastly, the fact that pulses of current are supplied by the transistor due to either limiting action causes distortion of the sinewave. The amount of distortion depends on the Q of the tuned circuit and on the loop gain of the oscillator. The greater the value of Q and the nearer the value of loop gain to unity, the better the quality of the sinewave.

The Two-winding Oscillator

Fig. 6.9 shows this type of oscillator in several different but equivalent forms. The tuned circuit may be on the input or on the output side of the active device, or it may be completely separate as far as d.c. is concerned, coupled by mutual inductance to other coils. In any of these forms, it is a four terminal oscillator. The differences between this oscillator and the blocking oscillator are important, as they illustrate how similar circuits can give different waveforms.

(1) The sinewave oscillator uses a tuned circuit in the feedback path.

(2) There is no large capacitor to charge up and down.

(3) The amount of feedback is much smaller because there is less coupling between the coils, so that the rise and fall of voltage is less rapid.

(4) The total gain (amplifier gain × feedback loss) is slightly more than unity, and drops to unity when the active device is driven into the non-linear region (when a valve takes grid current and a transistor takes excessive base current).

(5) The operating frequency of the oscillator depends much more on the tuned circuit, less on supply voltages.

(6) The quality of the sinewave depends on the Q of the tuned circuit *when it is loaded* (the Q is lowered by connection of a tuned circuit to other circuits).

Tapped Coil Oscillators

Fig. 6.10 shows a collection of oscillators of the type which may be classed as 'tapped coil'. These are really a form of the two-winding oscillator in which the two-winding transformer has been replaced by an auto-transformer with a tuned winding. In some cases, the actual tuned winding may be separate and inductively coupled to the tapped coil. These oscillators are classed as four-terminal since the autotransformer winding is doing the same job as two separate windings with four terminals.

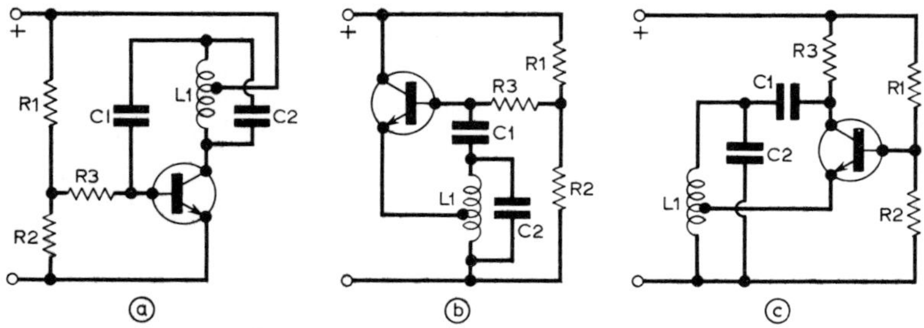

Fig. 6.10 Tapped coil oscillators.

Common-base Oscillators

This type of oscillator (Fig. 6.11) is very common in transistor circuits, and hardly fits into any of the previous classes. The output of a common-base-connected transistor is in phase with its input, and oscillation will occur when the collector is connected back to the emitter by a network. If a series tuned circuit is used, the feed-back is a maximum at the frequency of resonance, and so oscillation occurs at this frequency.

In general, any network connected between collector and emitter of a common-base transistor will cause oscillation at the frequency for which the network has minimum impedance or at which the trans tor has maximum gain. This common-base connection has also the advantage of allowing a transistor to oscillate at a higher frequency than would be the case with any other type of connection.

The four-terminal oscillators examined so far are not genuinely different types, as is shown later, although these and many others have distinct names (not given here) taken from their discoverers. The *Franklin Oscillator* is a two terminal oscillator, one of the few oscillator circuits using two active devices.

The circuit (Fig. 6.12) is that of a multivibrator with a tuned

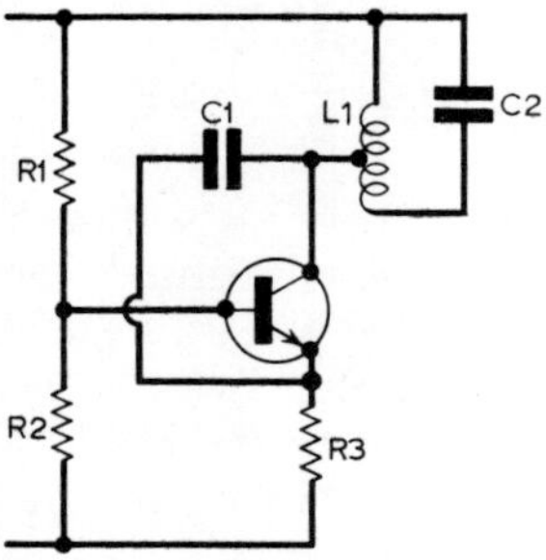

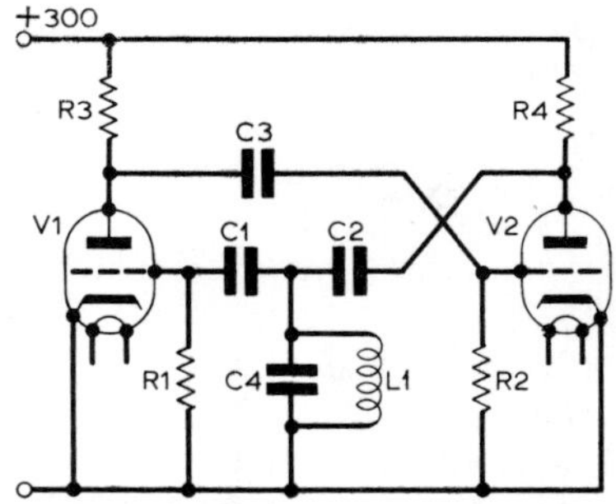

Fig. 6.11 (left) Common-base oscillator.

Fig. 6.12 (right) Franklin oscillator.

circuit forming a potential divider with a capacitor of low value. The capacitors *C*1 and *C*2 can be adjusted so that the positive feedback is very small except at the frequency of parallel resonance (usual values are about 1–2pF for *C*1 and *C*2). A crystal may be used instead of the *LC* circuit, and it is notable that crystals which do not work in any other circuit work very well in this case, because the crystal is loaded very little by the high impedance connections. Variations in capacitance in the remainder of the circuit have very little effect, and no tappings are required in the tuned circuit.

Op-Amp Oscillators

The action of an op-amp ensures that whenever a small voltage difference exists between the input terminals, the output is at either positive or negative line voltage. The high loop gain available with op-amps makes them highly suitable for generating non-sinewaves, though the slew rate limitations apply to the sides of square waveforms.

Two possible circuit forms exist, one using a steady bias on the inverting input and time-constant feedback to the non-inverting input, the other form being the exact opposite. Taking the second case—more often used for non-sine oscillation—first, Fig. 6.13 shows a multivibrator in which the + input is fed through a resistor divider from the output and the − input is connected to the *RC* network shown.

When the output is at positive line voltage, the + input is held at a positive voltage [+ line voltage × $R1/(R1 + R2)$], and the terminal B starts at − line voltage but is increasing positively as *C*1 charges up through *R*3. When the voltage at the − input comes within the range of offset voltage from the +input voltage, the op-amp output switches over, making the output at the −line voltage, and the +input at a negative voltage [−line × $R1/(R1-R2)$]. The cycle then repeats in the opposite direction.

The output waveform is a squarewave whose wavetimes at each level voltage are equal if the + and − line voltages are equal; the total time is given (in seconds) by $T = 4{\cdot}6\ CR \log(1+2R1/R2)$ where C and R are in Farads and ohms respectively. If C is in μF and R in kΩ, the time is in mS. This circuit can be modified to allow different times of charge and discharge by having two resistors feed the capacitor $C1$, each controlled by a diode so that one resistor charges and the other discharges.

The astable multivibrator described can be converted into a monostable circuit by the use of a diode across the capacitor to clamp the negative input to earth for positive pulses. A negative trigger pulse applied at the −input then flips over the amplifier and C then charges through $R3$ as before. When the output flips back again, the −input voltage is held at earth potential by the diode, preventing further oscillation. Bistable switches can be similarly produced by leaving the −input free and feeding a trigger pulse directly to it.

For sinewave oscillation, the loop gain must be limited, and must exceed unity only for the required oscillation frequency. For this requirement, it is better to keep the −input at a fixed bias and take the feedback through a frequency-selective network to the +input. The selective network may be an *LC* circuit, a Wien bridge, a twin T or any other circuit which produces zero phase shift at some value of frequency.

Testing Sinewave Oscillators

As we have seen, oscillators produce good sinewaves only when the feedback used is very slight, so that oscillation is just sustained and no more. This means that some caution must be used when testing oscillators, because connecting leads to, or even taking earthed objects near, an oscillator, can cause oscillation to stop.

When an oscillator is coupled to an amplifier stage, the output of the amplifier can be checked to see if sinewaves from the oscillator are present. If so, the oscillator need not be checked at all. If not, the output of the oscillator must be checked, and several methods exist of doing this.

If only a multirange meter is available, the current to the oscillator stage can be checked. This must normally be done by breaking the

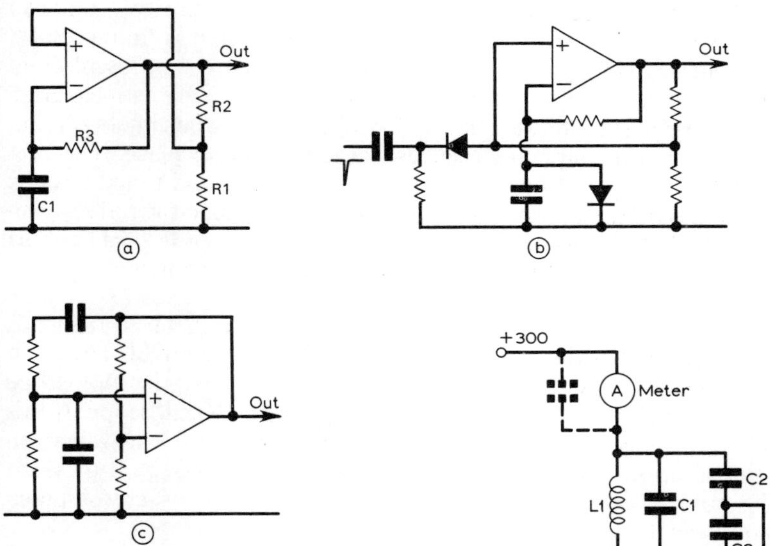

Fig. 6.13 Op-amp oscillators: (a) MV, free running. Power supplies not shown; (b) Conversion to triggered form by clamping diodes. (c) Sinewave oscillator using Wein Bridge. In practice, sinewaves would not be obtained unless a method of limiting gain (thermistor or FET) at peak amplitude, were used.

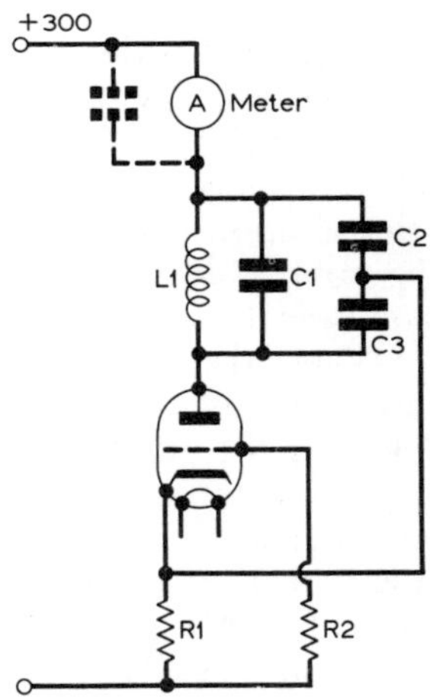

Fig. 6.14 Metering oscillator current. A large value capacitor may be connected across the meter as indicated.

power supply to the oscillator stage at a point where only the oscillator is supplied. (Fig. 6.14). The meter must *not* be connected in at any point where oscillating voltage exists, but only on the decoupled portion of the supply, sometimes called the cold, dead, or earthy end (as distinct from the hot, live, power end) of the circuit.

Alternatively the other side, cathode or emitter, may be broken so that a meter can be inserted. In some cases, the supply to the oscillator is filtered to prevent the oscillator output feeding forward to other stages (or having other voltages fed back), and the meter can be placed across one of the resistors to measure the *voltage* without breaking any leads; the current can then be calculated.

The current flowing in an oscillator is a very sensitive guide to the working of the oscillator, for when oscillation is taking place, the current is very much less than it is when there is no oscillation. If the tuned circuit is shorted while the current is being measured, the current should rise sharply, to drop again when the s/c is removed. If this happens, then oscillation is taking place. This test can be carried out on most oscillators using an *LC* tuned circuit; on some others it is necessary to use a large capacitor (large compared to any capacitor in the tuned circuit, perhaps 0·1μF) to short the tuned circuit to signal frequency without shorting out d.c. supplies.

In some cases it may be impossible to insert a meter in the leads without serious disturbance. If the oscillator uses an *LC* circuit, it is possible to detect the oscillation on a wide band oscilloscope using a pickup coil of one turn feeding into the oscilloscope. If this pickup coil is taken gradually near the oscillator coil, the waveform picked up should be visible on the oscilloscope, though it may not be possible to detect individual cycles. If the pickup coil is taken too near the oscillator, oscillation inevitably stops.

Where the oscillator uses a *CR* circuit, other methods must be used. An oscilloscope lead with a resistor probe can be used to monitor the output directly, the resistor (100kΩ or so) preventing the capacitance of the coupling link from affecting the output of the oscillator. Remember, if voltage readings are being taken, that the 100kΩ resistor forms a potential divider with the input resistance (usually 1MΩ–10MΩ) of the oscilloscope.

Crystal oscillators usually have an inductive load, and the oscillation can be detected either by current measurement or by pickup from the load inductor.

High power oscillators, used in industrial heating and ultrasonics generation, usually have built-in provision for measuring oscillator current. In the case of the large valve oscillators used in inductive or capacitive heating equipment (where an object to be heated is made the secondary of an output transformer if it is a metal, or the dielectric of a capacitor if it is an insulator) the grid 'resistor' is usually a set of lamp bulbs which light up while oscillation takes place.

CR Sinewave Oscillators

Fig. 6.15 shows a selection of these oscillators, all of which use some sort of complex *CR* networks. Phase shift oscillators use one amplifying stage, and shift the phase by 180° in the *CR* network (remember that 180° phase shift is the same as inversion for a sinewave only) so that the output of the network is in phase with signal at the input of the oscillating device. The minimum/maximum impedance networks such as the various 'T' circuits depend on the action of certain networks having either minimum impedance or maximum impedance.

A network which has minimum impedance is placed in a *positive feedback* loop so that oscillation takes place at the frequency of minimum impedance; a network which has *maximum* impedance is placed in a negative feedback loop so that oscillation is prevented except at the frequency of maximum impedance, or it may be used to short out a positive feedback loop except at the maximum impedance frequency.

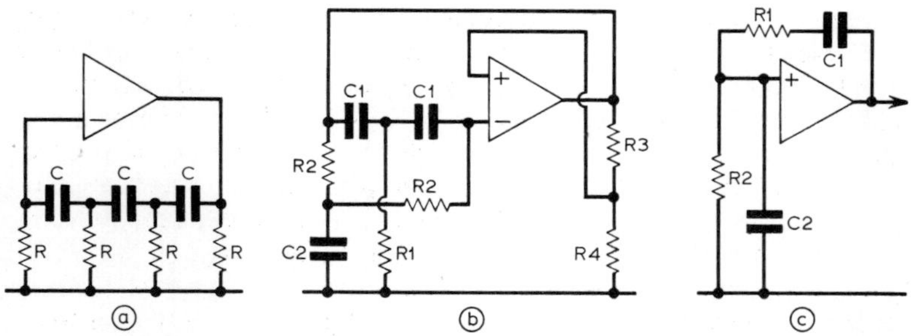

Fig. 6.15 Sinewave oscillators using C.R. networks: (a) Phase-shift oscillator. When equal values of R.C. are used oscillation is at the frequency where the phase shift of each R.C. is 60°; (b) Parallel-T network. If C2 = 2C1 and R2 = 2R1 then $f = 1/(2\pi C1 R2)$; (c) Wein Bridge. If C1 = C2 and R1 = R2, $f = 1/(2\pi C1 R1)$.

In general, *CR* networks have a very low *Q*, and some extra means has to be used to ensure that sinewaves are generated rather than squarewaves. In most cases, this takes the form of a device which increases negative feedback, or shorts out some of the positive feedback whenever the voltage across it or the current through it becomes excessive.

A thermistor is frequently used for this purpose; the resistance falls sharply as current or voltage (hence wattage and temperature) rise, and so the thermistor may be used as a load for the amplifier so that the output to the feedback loop drops when voltage or current becomes excessive. Other forms of non-linear resistors whose resistance increases with increasing voltage may be used in negative feedback loops; a lamp bulb in the cathode of a valve used to be a very popular method of controlling valve *CR* oscillators.

Crystal Control

Several types of crystal (such as quartz, a crystalline form of sand) have the peculiar property of vibrating mechanically when signal voltage is applied to metal contacts on opposite sides of the crystal. This vibration takes place at only one frequency (and some multiples, such as twice, three times etc.), and is very selective; frequencies only slightly different from the crystal frequency have practically no effect.

The crystal acts electrically as a *resonant circuit* of very high Q, and its electrical equivalent circuit is therefore of a resonant circuit with such very little resistance that the circuit could not possibly be constructed with any materials normally available to us. To make the point clearer, we can construct linear circuits with a Q of 300 by using the multistranded wire of the highest conductivity and avoiding losses as far as possible. A simple crystal can have a Q of 3,000 without any resort to any complex construction. For this reason, crystal oscillators are used (Fig. 6.16) wherever a very pure sinewave of very accurately determined frequency is needed: in transmitters, standard frequency supplies, certain receivers, etc.

In use, the crystal is treated exactly as if it were a tuned circuit, the only precaution being that the signal current through the crystal must be kept very low (to avoid shattering the crystal through excessive vibration). The crystal can behave as a *series* resonant

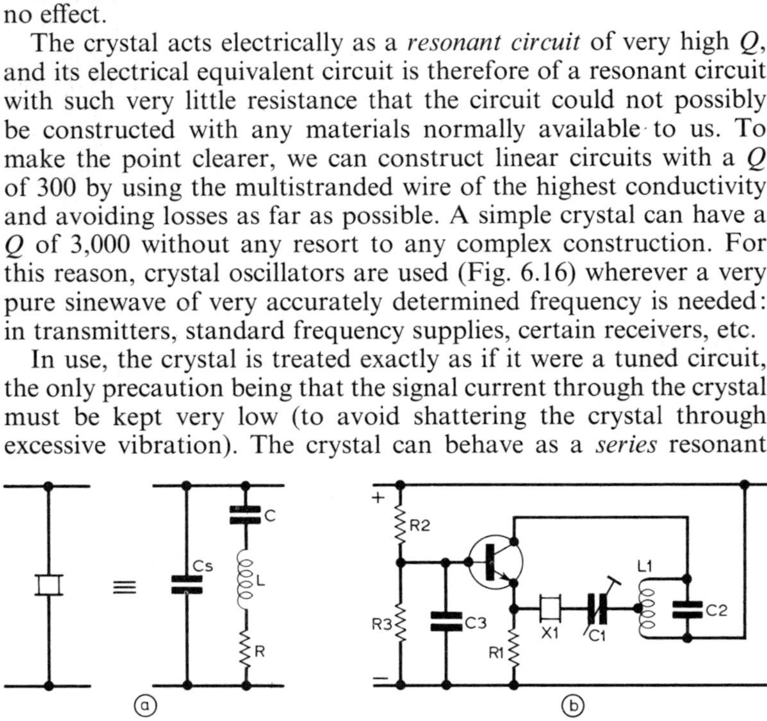

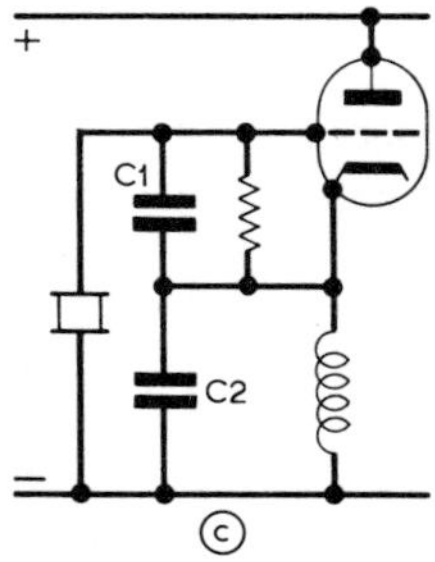

Fig. 6.16 Crystal oscillators. Most oscillator circuits will work when crystals are used; the examples shown are circuits not usually met with LC tuning. (a) Equivalent circuit of crystal. Cs is the stray capacitance across the crystal and its holder; (b) One form of transistor crystal oscillator. Note the use of common-base connection and the trimmer C1 for adjusting driving current. Excessive driving current can damage the crystal; (c) One form of triode crystal oscillator. In this form, the capacitors C1, C2 act to step up the signal voltage at the cathode so as to drive the grid, and the anode is left free.

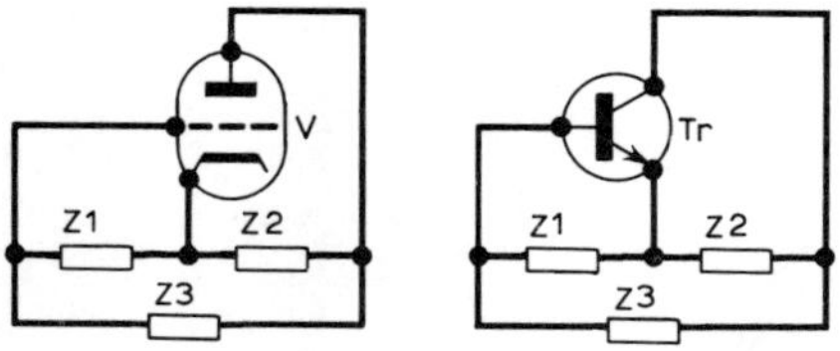

Fig. 6.17 General oscillator circuit, supply and bias lines omitted.

circuit or as a *parallel* resonant circuit; the oscillating frequencies for these two cases are slightly different.

Equivalent Circuit of an Oscillator

The classification of oscillators which we have used so far covers only the main types usually seen. In fact, practically all one-valve/transistor oscillators can be represented by the equivalent circuit of Fig. 6.17. This shows three impedances (*Z*1, *Z*2 and *Z*3) connected to a valve or transistor, and one condition for oscillation is that *Z*1 and *Z*2 should be alike (both capacitors or both inductors) and that *Z*3 should be an opposite reactance.

For example, if *Z*1 and *Z*2 are capacitors, *Z*3 is an inductor (a type of oscillator sometimes called *Colpitts*); if *Z*1 and *Z*2 are inductors, or a tapped inductor, *Z*3 is a capacitor (a type of oscillator sometimes called *Hartley*). From each of these basic types, three oscillating circuits can be constructed by making the earth point at different junctions. The theory of operation is unaffected by the position of earth, and so this may be placed at the junction of *Z*1 and *Z*2 (common cathode or emitter), at the junction of *Z*1 and *Z*2 (common grid or base) or at the junction of *Z*2 or *Z*3 (common anode or collector).

When these circuits are then redrawn in the conventional form, they appear to be quite different circuits, and this is especially true when crystals are used as the *Z*3 of the circuit, since the equivalent circuit of the crystal 'contains' a portion of *Z*1 or *Z*2 as well.

In addition to the circuits described, several devices contain both amplifying and feedback elements and are used as oscillators. Several semiconductor devices, such as Gunn diodes, tunnel diodes, electroluminescent diodes, give either a switching action or generate oscillations at very high frequencies. Electron beam devices with high power output of microwave energy include Klystrons and Magnetrons.

CHAPTER SEVEN

PULSE CIRCUITS

THE TYPES OF CIRCUITS used in radio communications are concerned with the transmission and amplification of waveforms with the least possible change in the shape of the waveforms. The other important branch of circuitry deals with pulse waveforms which are used for *switching* and for marking periods of time. For example, both in radar and in television we must use pulses, which repeat at regular intervals, to start off other waveforms, such as scans, so that these other waveforms are synchronised to any waveforms in other devices which are started by these same pulses.

The usefulness of pulses is that we no longer have to worry about distortion (except for time distortion) of waveforms, so that we can transmit or receive pulse signals under conditions which would make other communication impossible. The signal power of a pulse may be very much greater (because it is *concentrated* into one pulse) than the signal power of a sinewave, which is continuous.

We can perform operations of counting, storing, switching and measuring with pulses which would be very difficult with sinewaves; because pulses have sharp changes of voltage at precise intervals rather than the steadily changing voltages of sinewaves. Most important of all, we can generate pulses in an oscillator and *synchronise* them to other pulses so that the locally generated pulses occur always at the same time as the signal pulses.

Functions of Pulse Circuits

The types of circuit used with pulses differ considerably from those used for other waveforms. When we amplify pulses, we are not concerned with preserving the *shape* of the pulse, but with preserving the *time* at which it happens compared with other pulses. If our time standard is the *leading* edge of a pulse (if, for example, it is the leading edge of the pulse which starts off a timebase), then the leading edge must preserve its *rise-time* in the amplifier, but what happens to the rest of the pulse is immaterial.

We may wish to differentiate a pulse, to sharpen up the leading edge, or occasionally we may wish to *integrate* pulses so as to generate a voltage proportional to the width or number of pulses. We may wish to generate a squarewave of fixed or variable width from a pulse, or to generate a timebase waveform, or we may wish to use one pulse to open a gate to allow other waveforms through (to *count* them, for example) and another to close the gate.

We may wish to use a pulse to generate another at a fixed or variable time later, so that one waveform can be started before another (for example, the beam of an oscilloscope may be turned on just before the scan starts), or we may use a pulse as a comparison to force another waveform to occur at the same time as the pulse (as in automatic frequency controls). The pulse circuits which we shall now examine are the basic 'building bricks' of all pulse circuitry, and we shall also see how they can be inter-connected to carry out various tasks.

Pulse Amplifiers

If a pulse is applied to an amplifier intended for normal use, the results are disappointing. The amplifier is usually overloaded, and spreads out the pulse, the amplifier load is usually so high that only a small first portion of the pulse is amplified, and the rise time is greatly lengthened because the stray capacitance across the load resistor *integrates* the pulse. Changes in the d.c. level of the pulse affect the d.c. level of the output, and, worst of all, any voltage variations, noise, etc., are amplified to the same extent as the pulse.

By contrast, a pulse amplifier is shown in Fig. 7.1. This amplifier is intended for use with positive-going input pulses, and is normally

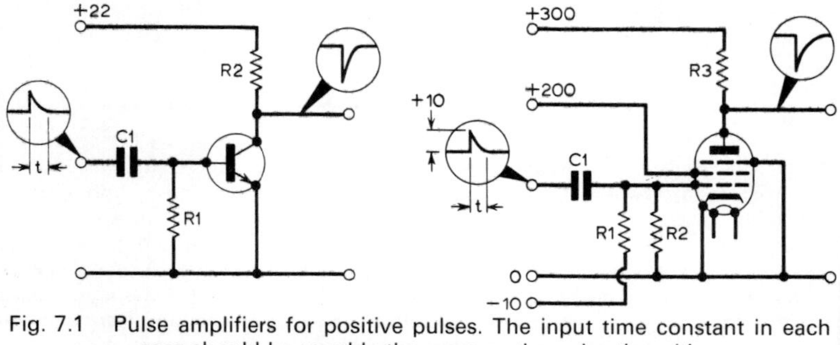

Fig. 7.1 Pulse amplifiers for positive pulses. The input time constant in each case should be roughly the same as the pulse time, 't'.

biased off so that no current flows. This means that stray voltages and noise, as long as they are not the same voltage as the pulse, have no effect since the bias can be arranged so that such spurious signals do not turn the amplifier on.

The next point is that, by using the input pulse to turn current in the valve or transistor *on*, the *valve* or *transistor* is supplying the current to charge up the stray capacitances (a 25pF capacitor charged up 10V in 100nS requires a current of 2·5mA simply for charging, and in addition to any current passed through the load resistor); this can usually be provided. If, on the other hand, the valve or transistor were being shut *off*, the voltage at the output would drop at a rate determined by the time constant of load resistor and stray capacitance (so that voltage moved by 2/3 of the range in time CR), and this shape of voltage waveform would not be the same as that of the input pulse.

By selecting the bias, we can choose to amplify only a tip of the input pulse, or we can clip the top of the pulse with a diode and amplify some selected portion lower down. If the pulse has a sloping side, we can introduce some variation in the *time* of the output pulse, by doing this.

It is possible, too, to increase the immunity of a pulse circuit to noise by 'gating'—turning the amplifier on only when a pulse is expected. Since most pulse circuits work with short duration pulses which repeat at fixed (and much longer in time scale) intervals, the time of arrival of a pulse is predictable.

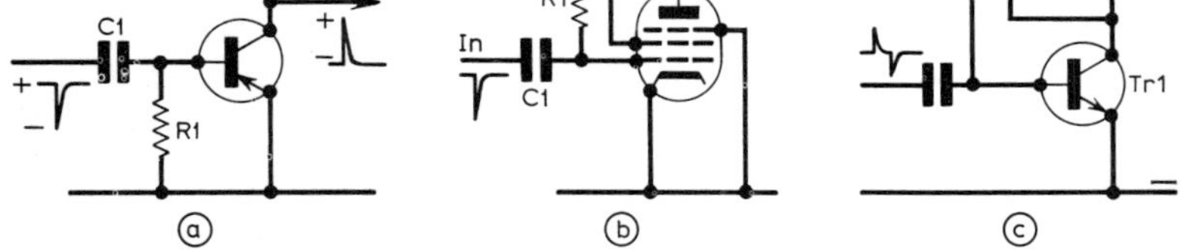

Fig. 7.2 Pulse amplifiers for negative pulses: (a) transistor version, using p-n-p transistor; (b) Valve version, using pentode held in conduction by grid return to positive voltage. R1 should be a high value, 0·5MΩ or more, to avoid excessive grid current. The value of R2 will determine the dissipation in the valve and the rise-time of the pulse out; (c) Double output stage. Taking the output as shown from the emitter is an advantage if the positive rise time is the more important; if the negative rise time is more important the output can be taken from the collector of Tr1.

Fig. 7.2 shows amplifiers for use with negative-going input pulses. The use of a p-n-p transistor is the easiest solution, because a p-n-p transistor is switched *on* by a negative-going input pulse, and the action is exactly as described for the case of a positive-going pulse with a n-p-n transistor.

If only valves, or transistors of one polarity type, are to be used in an amplifier, the problem of dealing with pulses which switch the amplifier *off* has to be faced. A very large standing current may be passed through a low load, so that the time constant at the output of the amplifier is small. Much better are the arrangements of Fig. 7.2(c), where the shut-off of current in one device causes current to be turned on in another, so forcing the output voltage to change sharply. This type of circuit is widely used under a variety of names.

Pulse Shapers

When a pulse amplifier is used together with differentiating or integrating circuits, the complete circuit is known as a pulse shaper. In addition to differentiation or integration, clipping and level setting may also be carried out in a pulse shaper. One common use for pulse shaping is to sharpen up the rounded pulse obtained from the receiver of a radar set. Similarly, the trigger amplifier of an oscilloscope has to provide a sharp pulse output from an input which may be of any waveform, and in this case a considerable degree of shaping and amplification may be needed. In some cases, a *monostable* may be used to create a squarewave from any input which is large enough to trigger the monostable.

Monostables

A monostable is a relaxation oscillator in which the active components are permanently biased in one state (one on, one off) and which does not generate continuous oscillations.

An input signal which is *not* of sufficient amplitude to turn the stage to which it is fed to the reverse state (on, if it were off: off, if it were on) has *no* effect on a monostable. There is no output. If, however, the input signal exceeds the critical value, called the *trigger threshold*, the monostable changes over (flips) rapidly. Due to the time constant coupling, the *flip* state does not last, but eventually reverses to the original condition (*flops*) equally rapidly.

The output of the monostable, or *flip-flop* is as shown in Fig. 7.3, together with typical circuits. It consists of a square pulse whose width is set by the time constant used, and the voltage to which the bias resistor of the *on* transistor is returned. This pulse width can be made *variable* by making the R of the CR a variable resistor, or by switching the value of the capacitors.

The monostable can therefore be used to generate a pulse of varying width from a given input. An input suitable for operating a monostable is termed a *trigger*, and ideally should be a very narrow spike of voltage of the correct polarity. The monostable will always be driven by the trigger; if the frequency of the trigger pulses is increased, the frequency of the output pulses from the monostable is also increased until the situation arises when one output squarewave overlaps another (when trigger spacing in microseconds equals squarewave width), when the monostable is switched permanently over in the other direction.

A monostable can also be used to generate a *time delay*, and the block diagram of a method of doing this is shown in Fig. 7.3(c). The block diagram of the time delay shows a trigger amplifier, a monostable, a differentiator and a diode pulse selector; it shows also the waveforms to be expected at each junction. The trigger pulse from the trigger amplifier operates the monostable in the usual way, and a squarewave is produced at the output of the monostable.

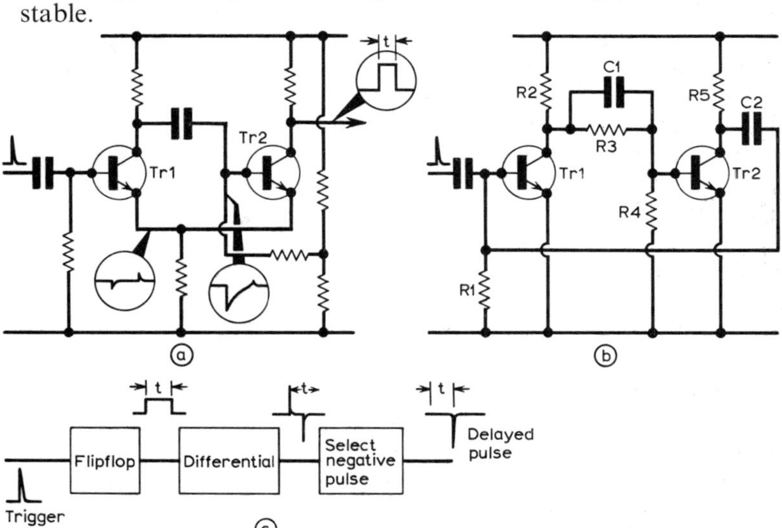

Fig. 7.3 (a) Emitter-coupled flip-flop. There is no output unless the trigger pulse is of an amplitude great enough to cause Tr1 to conduct; (b) Cross-coupled flip-flop. The small capacitor C1 ensures fast leading edges by speeding up the transition of Tr2 from 'on' to 'off'; (c) Obtaining a time delay. By variation of the time constant of the flip-flop, the delay can be made variable.

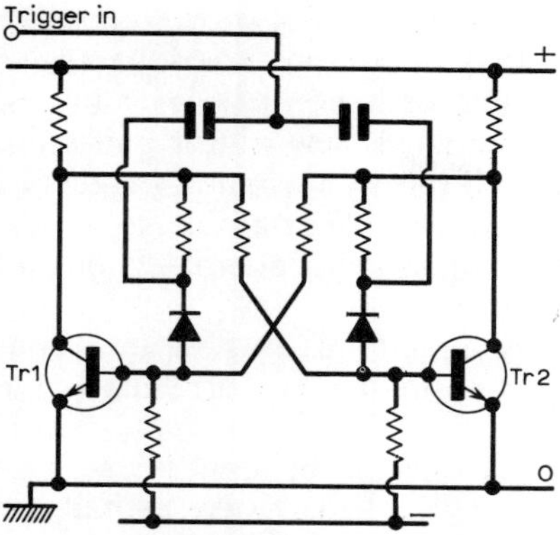

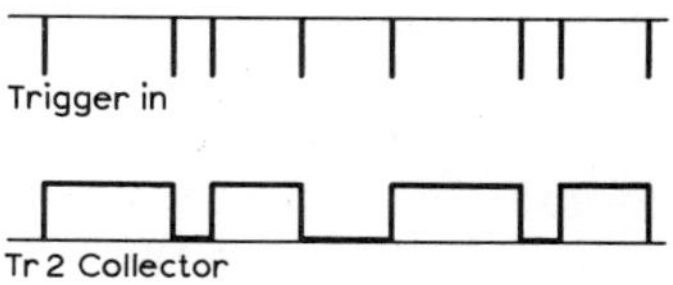

Fig. 7.4 Bistable. One transistor is conducting, with its base held on by the collector voltage of the other transistor. When a negative-going trigger arrives, the base which is more positive is switched off, and the other transistor conducts, switching the first one off. The diodes, called steering diodes, ensure that the trigger affects only the base which is more positive. The circuit changes over each time a trigger of suitable amplitude is received.

This squarewave is differentiated to produce two spikes—one for the leading edge and the other for the trailing edge, and the leading edge spike is removed by a diode. The spike output is a pulse which always occurs a set *time* after the input pulse, and this time may be varied as before by altering the time constants or the bias voltage of the monostable (providing the pulse frequency does not cause overlap), but once set will always follow the trigger pulse.

Variation on the basic monostable include diodes to set exactly the bias and catching levels, the use of cathode/emitter follower between the sections of the monostable, and provision for feeding a 'switch-off' pulse into the other grid/base so that the monostable can be made to give a shorter pulse if the switch-off pulse is received just after the normal trigger pulse. A blocking oscillator may also be made monostable by biasing off the grid/base so that continuous oscillation does not take place.

Bistables

Whereas a monostable has only one normal condition, which can be switched over by a trigger pulse, but which always switches back again; a bistable has two 'normal' states. Each trigger pulse received will cause the bistable to change over, but in the absence of a trigger pulse there is no further change. Fig. 7.4 shows a bistable circuit and its waveforms. Note that the output is a squarewave whose wave time is equal to the time between *two* trigger pulses. If the output of a bistable is differentiated, new trigger pulses are generated which occur at half of the rate of incoming triggers of the same polarity. The bistable has several important uses:

(1) It can act as a switching circuit; it switches alternately on and off each time a trigger pulse is received.

(2) It can be used as a counting circuit; it can give an output trigger for each two input triggers no matter how irregularly the input triggers are spaced.

(3) It can be used as a frequency divider; if the input is a squarewave differentiated, then the output is a squarewave at half the frequency of the input.

(4) It can be used as a memory store; when a bistable receives a trigger pulse it flips over and remains in that state until another trigger pulse is received.

The Schmitt Trigger

The Schmitt (or Schmidt) trigger is a form of emitter-coupled bistable in which the input signal is fed always to one base instead of being guided to alternate bases by diodes as is the case in the conventional bistable. The Schmitt trigger of Fig. 7.5 will switch over when the voltage at the 'in' terminal has reached a level set by the design of the trigger, and will switch back again when the voltage drops below another slightly lower level.

These voltage levels are called the Upper and Lower Threshold voltages, and the output of the Schmitt trigger consists of a step of voltage each time the input rises above the upper threshold or falls below the lower one. If a waveform is fed into the Schmitt trigger, providing that its amplitude is greater than the difference between the thresholds, then the output will be a squarewave of the same frequency. The Schmitt trigger is very widely used in pulse circuits, particularly in control systems and computing circuits. Its main uses are:

(1) As a converter from sine (or other) waveforms to square waves.

(2) As a voltage level detector—since it gives a step of voltage whenever the threshold voltage is crossed.

(3) As a form of relay, turning on when the input signal is high and turning off when it is low, or the other way round.

The difference between upper and lower threshold voltages can be increased or decreased by modifications to the design of the circuit as required.

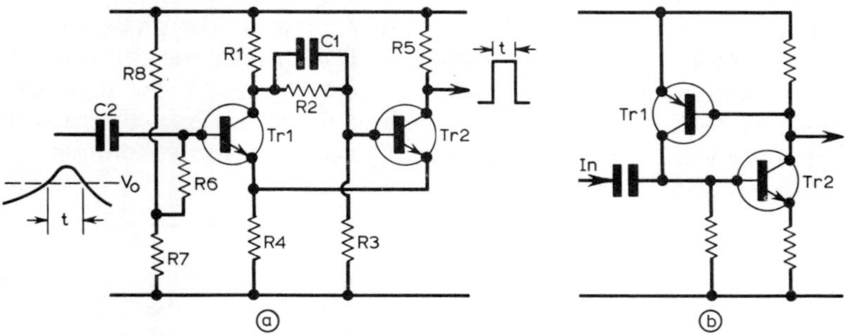

Fig. 7.5 Schmitt trigger circuits: (a) Vo is the voltage level at which Tr1, which is normally off, conducts. At this level, the circuit switches over, and does not switch back until the voltage at the base of Tr1 has dropped again, usually to a value rather lower than Vo; (b) This version gives a negative pulse at the output for a small positive voltage at the input. When Tr2 starts to conduct, Tr1 is switched on and increases the base current into Tr2, so speeding the switch-on. A similar action takes place on switchoff.

Gas Trigger Tubes

Gas filled tubes similar in design to gas stabilisers can also be used as trigger generators and switches. If a third electrode is inserted into a gas stabiliser, so that the gap between the cathode and the third (trigger) electrode is very small, then a positive voltage pulse applied to the trigger electrode can ionise the gas and start conduction, a process which will then continue if the voltage between anode and cathode is adequate.

The voltage required for triggering is much less than would have to be applied between anode and cathode to cause conduction. Once conducting, the tube remains conducting until the anode voltage is lowered to the point where the gas becomes de-ionised, the extinction voltage. Unlike the other devices described so far, reverse triggers or second triggers do not reverse the action; this form of action is sometimes called 'latching'.

Resetting is usually carried out by means of a reset pulse which momentarily lowers the anode voltage or raises the cathode voltage. This resetting may be automatic (as when it is obtained from a later stage in the form of a pulse) or manual, when a 'reset' switch is used.

Solid-state devices which give the same sort of latching action are the silicon-controlled switch and its close relative, the thyristor.

SCS devices are used at lower power levels, frequently in logic switching circuits; thyristors in higher power circuits particularly where control of a.c. operated devices is concerned. Control circuits require a pulse whose phase relative to the phase of the mains can be varied and which is used to trigger the thyristor at some point in the input waveform. Once triggered, the thyristor remains conducting until the voltage across it reverses or reaches too low a value to sustain the current passing above some minimum value.

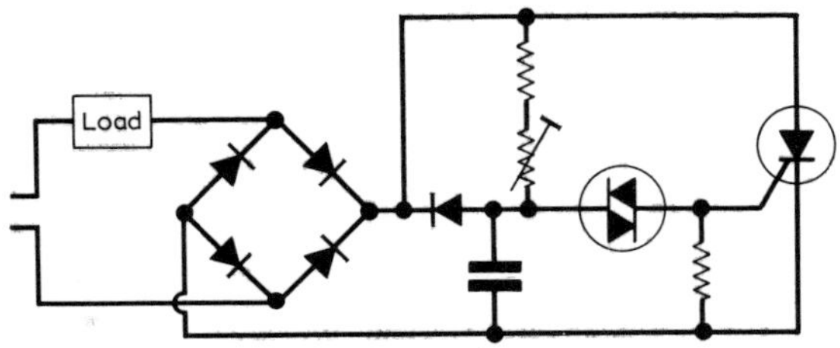

Fig. 7.6 Control of a.c. load using single thyristor and bridge rectifier.

As the thyristor is a controlled diode, it controls current over half a wave of a.c. only, and another thyristor must be used to control the other half; alternatively, one thyristor may be used in a bridge circuit as shown in Fig. 7.6, where the voltage applied to the thyristor is in one direction only. This circuit is often found as a method of controlling low power a.c. devices such as induction motors or fluorescent lights. The triggering circuit may include a four-layer diode which passes current suddenly at some value of forward voltage, so ensuring a definite and sharp trigger current to the gate of the thyristor.

Note that it is the current signal which is important for triggering, so that a fairly low impedance trigger source is needed. Some care has to be taken in the bridge circuit shown to ensure that the voltage across the thyristor falls to zero twice per cycle; any capacitance across the bridge will tend to prevent this, so that the thyristor remains on.

Thyristor Circuits

Another method of control is to use thyristors directly in series with the load, as shown in Fig. 7.7. In this circuit, control over one half of the cycle is dispensed with and only one thyristor used. The trigger diode, as before, breaks down at some definite voltage and produces a pulse in either direction of voltage. The trigger pulse may be applied directly to the gate of the thyristor or through a transformer, so avoiding d.c. level difficulties.

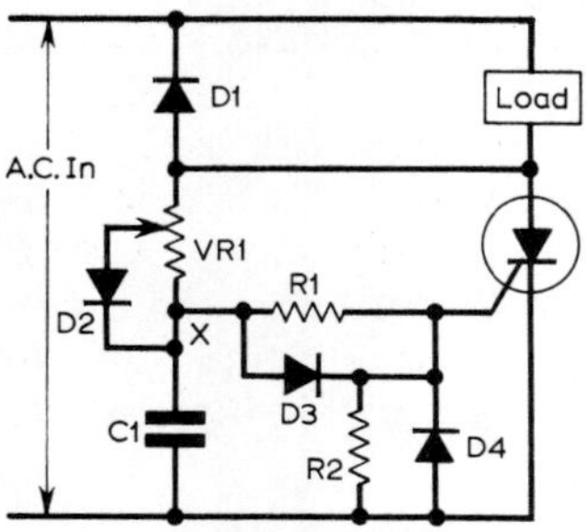

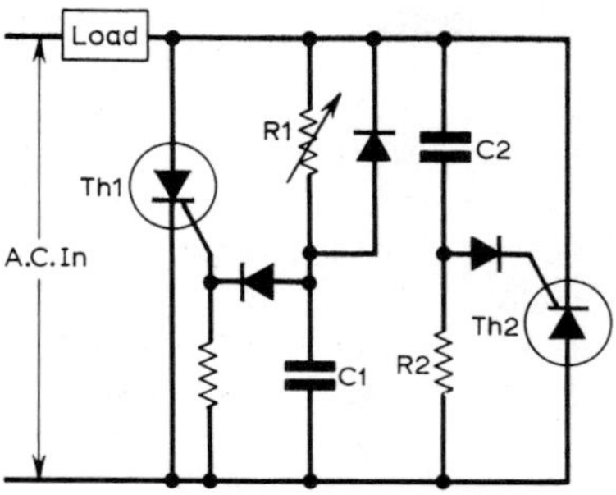

Fig. 7.7 (left) Series load operation. This circuit is commonly used for motor control. The speed with which the voltage at point X reaches the triggering level of the diode D4 depends on the size of the capacitor C1 and the setting of VR1. D1 absorbs voltage surges in an inductive load which might otherwise cause unwanted triggering.

Fig. 7.8 (right) Full-wave operation. This circuit uses a second thyristor whose firing point is controlled by the first thyristor as explained in the text.

Another method of full-wave operation is shown in Fig. 7.8, where the firing phase of the second thyristor depends on the previous firing of the first. If the first thyristor fires early in the cycle, the voltage across it is low, and this is the voltage to which the capacitor *C*2 charges. When the voltage reverses, *C*2 charges through *R*2 and fires the thyristor when the gate voltage is sufficiently positive to cathode. This will happen at a fairly low voltage if *C*2 started with a small negative bias. If, however, the first thyristor had not conducted, *C*2 would have charged to the full peak voltage, holding the gate of the second thyristor firmly negative throughout the second half of the cycle. Some care has to be taken that the reverse voltage rating of the diode in series with the gate is adequate to stand the peak reverse voltage.

Thyristors can be used as latching relays, and a huge variety of circuits has been devised to perform the essential switch-off operation. One basic circuit uses a pair of thyristors as a circuit breaker (Fig. 7.9). In normal operation, *Th*1 conducts on receiving a pulse from the START switch. The voltage across the load charges the capacitor *C*1 and the voltage across the resistor *R*2 in series with *Th*1 is applied to the four-layer diode in the trigger circuit of *Th*2. If the load current becomes excessive, *Th*2 is turned on, connecting the positively charged plate of *C*1 to the *cathode* of *Th*1 and so turning it off. By the time that *C*1 has discharged, *Th*1 is off, and the circuit can be reset. The action of resetting turns off *Th*2. Note that this is a d.c. circuit.

Thyristors can also be used in high power timebases and oscillators. Basically, they are used to switch current in inductors, and the back-e.m.f. or autotransformer action is used to turn off one thyristor when another fires. Timebases are dealt with later. Thyristor oscillators are met with in inverter circuits where d.c. is converted to a.c. with high efficiency and at high output powers. There is by now a large mass of circuit information on the use of thyristors, and this brief description is intended only as a guide to the basic uses, since an exhaustive treatment of this topic would fill this and several other volumes.

Failure of Pulse Amplifiers, Monostables, Bistables

In every case where a pulse operated circuit has failed, meaning that no signal output is detectable with an oscilloscope set to the correct ranges of voltage and time, and with the correct input applied, the following checks should be made, remembering that narrow pulses can be very difficult to detect on a 'scope unless the setting up is exact.

(1) Check that the input pulse is of the correct peak voltage and width. It is good practice to keep a note of all waveforms, with voltages and times shown, used as the input pulses to any piece of equipment.

(2) If the correct input conditions exist, check at the input of the first active device that the bias has not changed sufficiently to keep the device switched off (if the pulses switch it on) or on (if the pulses switch it off).

(3) Check that a power supply is available at the output of the first active device, and check also for signal at this point. If there is no signal, but bias and power supply are correct, then the first active device has failed, and the failure will have to be investigated, as it indicates either overloading or inadequate design.

(4) Repeat this procedure for each active device stage of the circuit.

A logical sequence of testing can save a great deal of time in checking pulse circuits. Where all-transistor or integrated circuits are used, failure of active devices in correctly designed and operated equipment is rare, and resistor failure should be suspected and investigated before any attempts are made to remove active devices. Valve circuits have the advantage in this respect because they are easily removable (although wired-in valves are found) and are not likely to be damaged by removal, so making a post-mortem easier than with transistors.

If the output of a stage exists, but is defective in voltage, then bias and power supply voltages should be checked. If the output pulse has changed in time (wider, narrower, differentiated, etc) from the normal, then time constant components should be examined.

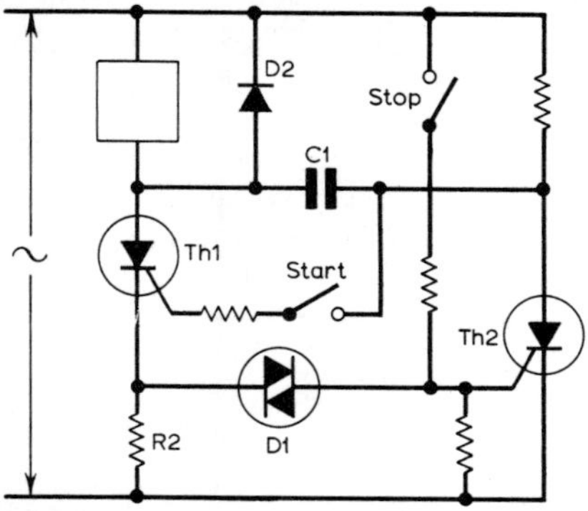

Fig. 7.9 Example of turn-off by a second thyristor. R2 is a small resistor whose value is such that excessive current in the load gives a voltage across R2 which is just enough to trigger the diode D1. Alternatively, triggering can be carried out by the 'STOP' switch.

Gating, Clamping and Switching

These three terms are closely related; all play a most important part in pulse circuits. An ideal switch has a very high impedance to signals in the 'off' position, and a very low impedance in the 'on' position. Mechanical switches are good in these respects at low speeds of operation, but at high speeds some form of electronic device must be used for switching, as electrons can be moved much more easily than the mechanical parts of a switch.

Transistors have a fairly high impedance when switched off (base back-biased) and a low impedance when switched on (base taking maximum current). For many electronic purposes, the low 'on' impedance is more important than a high 'off' impedance, and transistors make much better switches than valves in this respect. Like valves, transistors are bidirectional switches, meaning that they can be switched on or off, unlike gas triggers, thyratrons and other devices which can only be switched *on*, and which need different treatment to turn them off.

Any circuit which turns a transistor alternately fully on and then fully off could be called a switching circuit; the pulse amplifier of Fig. 7.1 is a switching circuit controlled by its input. Normally, however, we talk of a switching circuit only when one circuit is being used to control another, with no connection other than the switch.

Fig. 7.10 shows a gate controlled by a switching circuit. A gate circuit transmits a waveform through the gate when the gate is switched on by a pulse; when the pulse is absent nothing passes through the gate. There is a wide variety of gating circuits, but the principle is the same in each case.

Fig. 7.11 shows some time-coincidence circuits; these give an output when two inputs are received and although in this sense they might appear to be identical to gate circuits, there is an important

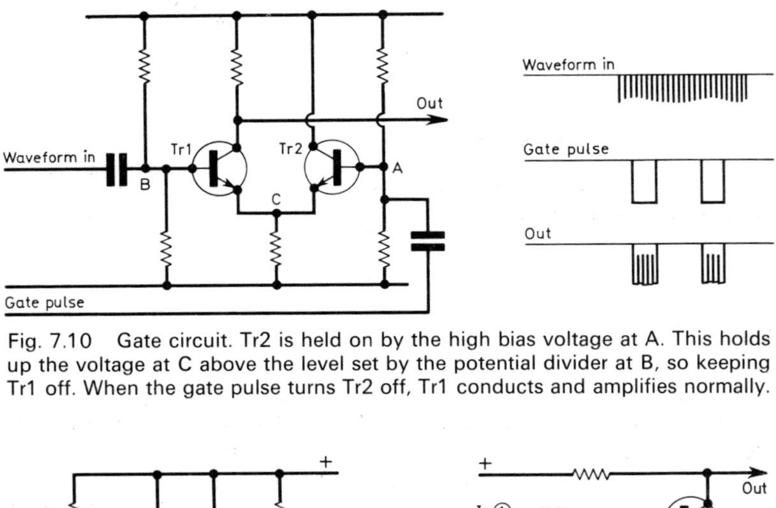

Fig. 7.10 Gate circuit. Tr2 is held on by the high bias voltage at A. This holds up the voltage at C above the level set by the potential divider at B, so keeping Tr1 off. When the gate pulse turns Tr2 off, Tr1 conducts and amplifies normally.

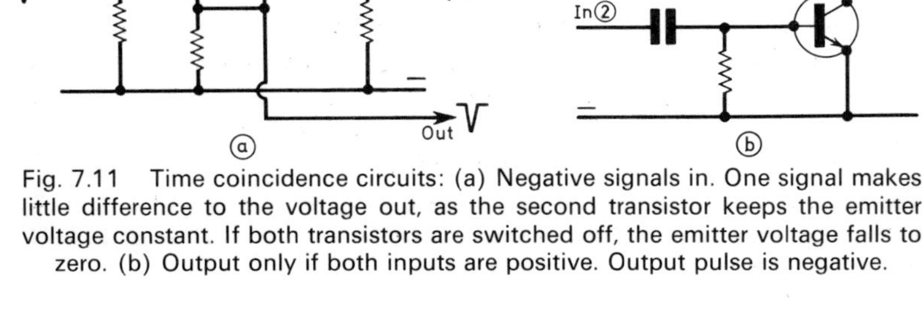

Fig. 7.11 Time coincidence circuits: (a) Negative signals in. One signal makes little difference to the voltage out, as the second transistor keeps the emitter voltage constant. If both transistors are switched off, the emitter voltage falls to zero. (b) Output only if both inputs are positive. Output pulse is negative.

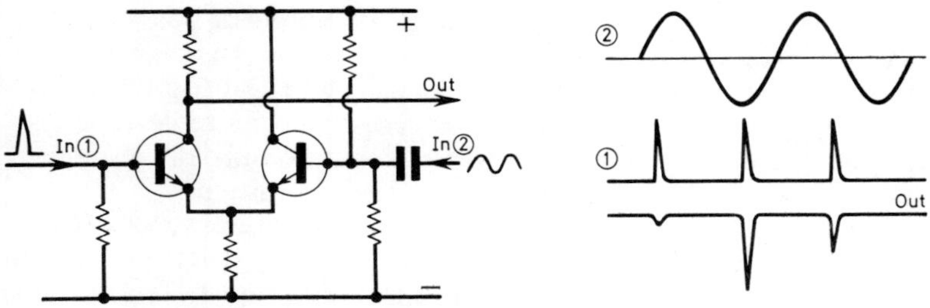

Fig. 7.12 Phase coincidence detector and waveforms.

difference. The gate circuit is intended to transmit a waveform faithfully, without avoidable distortion; the time-coincidence circuit simply gives an output of any shape not necessarily related to either input, whenever the two inputs coincide. Voltage-coincidence circuits give an output when a signal reaches a set voltage level; the Schmitt trigger is a good example of a voltage-coincidence circuit.

Phase-coincidence Circuits

The time-coincidence gate mentioned above is worth further study, as it may also be used for phase-coincidence detection. An example is shown in Fig. 7.12. The input to one transistor is a positive-going waveform which causes a large pulse output at the collector if the emitter voltage is low, but produces very little output if the emitter voltage is high, as the transistor is conducting on the tip of the pulse only.

The emitter voltage is controlled by emitter-coupling a second transistor which is fed with a sinewave (in this example) of the same frequency. When the positive peak of the sinewave coincides with the positive pulse, the emitter voltage is high and the output is low. When the negative peak of the sinewave coincides with the positive pulse, the emitter is low when the pulse occurs and the output is high. The output at the first collector may be integrated and used directly as a phase control voltage.

Timebases

Since pulse circuits are concerned with time rather than waveshape, circuits which can perform time measurement are an essential part of most pulse circuitry; such circuits are timebases. A timebase has as its input a trigger pulse, and as its output it has a voltage which rises or falls linearly with time and then returns sharply to its starting voltage. Such a voltage is the sawtooth of Fig. 7.13. The charging of a capacitor through a resistor is a simple timebase circuit; the output is not very linear, but rounds off at higher voltages. The blocking oscillator gives such an output.

Timebases are used in oscilloscopes, for measuring time between waveforms; in radar, for measuring distance by the time a wave takes to travel to and from a target; in other circuits for converting voltage signals into time signals by measuring how long the timebase output takes to reach the voltage of the waveform being measured.

Every timebase consists of three portions: a squarewave generator, a switching circuit and an integrator. As we should now be familiar with these separate portions, we can examine the combination used as a timebase. For most purposes we shall concentrate on the switching and integrating portions, since these are of greatest interest in forming the linear rundown or runup of voltage (or current) which is the output of the timebase.

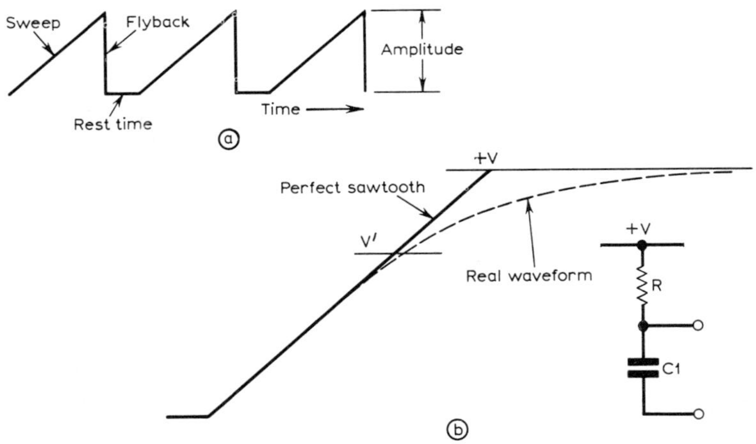

Fig. 7.13 Sawtooth timebase waveform: (a) Ideal waveform, (b) Waveform obtainable by capacitor charging. Time to reach about 2/3V is R.C. where R is in MΩ and C in μF, giving time in seconds.

The Simple CR Timebase

Fig. 7.14(a) shows the simple CR timebase. During the time of the pulse, the transistor is 'on' and the voltage at the collector is almost at earth. During the 'off' time, the voltage at the collector rises

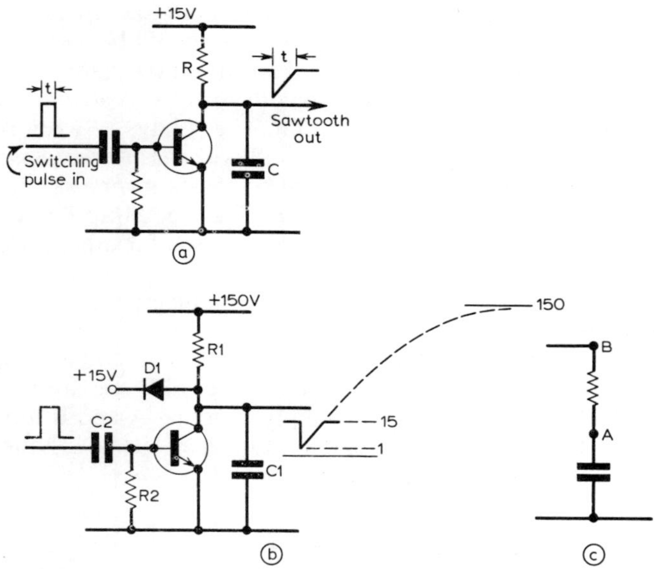

Fig. 7.14 Sawtooth generators: (a) Simple generator. The slope of the sawtooth depends on C.R., and the linearity will be poor unless only a volt or so is used; (b) Another way of using only a small part of the charging waveform. The capacitor is charging towards 150V, but is held at 15V by the diode. R1 must be fairly large to avoid excessive currents through D1.

steadily in the exponential charging curve until the capacitor has charged to line voltage or until another pulse discharges the capacitor to earth potential again.

The output is not linear, unless only a very small portion is used, or unless the charging resistor can be returned to a very high voltage, as in Fig. 7.14(b) where the transistor is protected from damage by a catching diode. For any sort of accurate measurement, more linear timebases are required and these require better integration.

Bootstrapping

Bootstrapping, as we have seen earlier, is a circuit technique which uses positive feedback. Referring back to the simple circuit, the reason that the runup of voltage is non-linear is that the charging current into the capacitor is not constant: it reduces as the capacitor charges, because the voltage across the resistor is decreasing as the capacitor voltage approaches the line voltage. If we could keep the current flowing through the resistor constant, the charging rate would be constant and the timebase output would be linear. Bootstrapping is one method of keeping this current constant.

Suppose (Fig. 7.14(c)) that as the voltage at 'A' rose, the voltage at 'B' rose by exactly the same amount. The voltage between A and B would then be constant although both voltages were rising, and the current through the resistor would be constant, since a constant voltage existed across it and $I = V/R$. The bootstrap circuit is a method of carrying out this process, and is used for other purposes where a constant current is required; for example to make a low impedance behave like a high impedance (which passes an almost constant current).

Fig. 7.15 shows an outline of a bootstrap circuit and a complete circuit. It is difficult to ensure that the voltage at B rises at the same rate as A at all frequencies, and simply coupling A and B together is no answer as it merely shorts them together. Using an amplifier

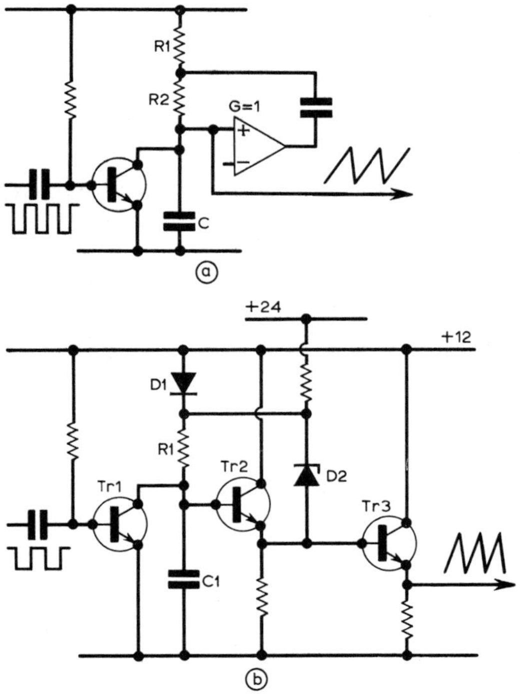

Fig. 7.15 Bootstrapping: (a) Principle of operation; (b) Practical circuit. The voltage across R1 is set by the zener diode D2, which must be returned to a higher voltage than that of the main supply. Tr3 ensures that the bootstrap circuit is not excessively loaded by the circuit connected to it.

with a gain of +1 (no inversion), point B can be driven by the voltage at A, and a much more linear timebase obtained. Because of the capacitor which couples the amplifier to B the voltage at B rises above the line voltage during the timebase sweep.

To avoid the effect of current flowing back to the line, the coupling from line to B can be made with a diode which cuts off when B is above line voltage. Note that the use of bootstrapping can be quite independent of any means used to discharge the capacitor, and of any squarewave generator. Though an emitter follower is popularly used in the bootstrap circuit, an amplifier of more stages can be used, provided that its gain is +1 and there is no inversion; for example, a long-tailed pair could be used.

As a timebase, the bootstrap has the disadvantage that the load on the charging capacitor must be very high; the circuit works best with a valve or a MOSFET as the amplifier stage. The linearity of the sweep depends on the gain of this amplifier and can be adjusted if the gain of the amplifier can be adjusted. The time taken to reach a voltage V is given by $t = V/V_L \times R \times C$ where V_L is line voltage in volts, R is the charging resistor, C is the charging capacitor.

If C is in μF and R in MΩ, t is in seconds.

If C is in nF and R in MΩ, t is in mS or C in μF, R in kΩ.

If C is in pF and R in MΩ, t is in μS.

Example: Line voltage is +20V, amplitude of sawtooth is 15V. What is time of sweep if $R = 10\text{k}\Omega$, $C = 1\mu\text{F}$? ($t = 15/20 \times 10 \times 1 =$ 7·5mS).

The Miller Timebase

This timebase depends on negative feedback for its action; a simple example is shown in Fig. 7.16. The charging capacitor is connected so that it is in a negative feedback loop in an inverting amplifier, and a resistor is connected so that the input of the amplifier can be connected to a switching circuit. Imagine the circuit at rest with the amplifier cut off, output voltage equal to line voltage and the capacitor fully charged.

If the charging resistor is now connected to a voltage sufficient to turn the amplifier fully on, a current flows through the resistor to charge the capacitor. Normally, the current through the resistor would become less as the voltage changed, but the negative feedback through the capacitor forces the current to remain steady although the capacitor is discharging all the time.

The sweep remains fairly linear until the amplifier bottoms, and non-linearity of the amplifier does not make the sweep non-linear.

The output impedance is low, so that connections may readily be made to the output, and, like the bootstrap, improvement can be made by adding extra stages. The greater the amplification inside the feedback loop, the more linear the sweep.

A characteristic of Miller timebases is the step of voltage at the beginning of the sweep (the Miller step) which is caused by the rapid growth of current in the charging resistor before the amplifier switches on and negative feedback starts. In the simple circuit, the sweep process is reversed when the Miller stage is switched off again, so that the flyback is as long as the sweep unless an extra switching stage is added to charge the capacitor rapidly by by-passing the load resistor (Fig. 7.17)

The Bootstrap circuit and the Miller circuit are two methods of ensuring linearity in a timebase, but they are not the only ways. Transistors and pentode valves are devices which both pass a steady current despite varying anode/collector voltages, providing their bias is constant, and for this reason they may be used as the charging resistors for a timebase.

This scheme works particularly well with transistors, because a mixture of p-n-p and n-p-n types can be used. Fig. 7.18 shows a circuit of this type used in a marine radar set (Decca Radar Ltd.) in which Zener diodes set the bias on one transistor, the current is varied by altering the emitter resistor, and the capacitor is discharged by a transistor of opposite polarity.

Inductive Timebases

A timebase may be generated by feeding a valve or transistor with a squarewave, using a high-value inductor for the anode or collector load. This circuit is used in television sets but seldom anywhere else, since control of the timebase linearity and speed is difficult. For further information on this subject, readers are referred to *TV Technician's Bench Manual*, published by Fountain Press.

Fault-finding in Timebases

Although all timebases basically depend on the charging of a capacitor or the growth of current in an inductor, they can be extremely complex in their circuitry, especially when requirements of high accuracy call for variations on the bootstrap or Miller circuits. For this reason, fault finding should be carried out keeping to a logical procedure.

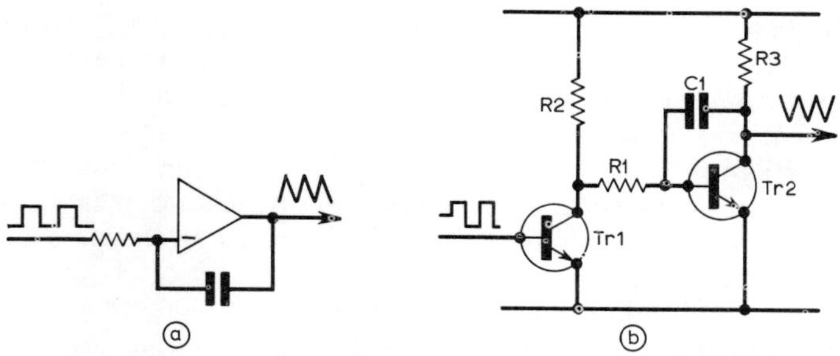

Fig. 7.16 Miller Integrator: (a) Working principle, (b) Practical circuit. When Tr1 is shut off, C1 charges through R1, giving a negative slope output of long time constant C1 (R1 + R2). When Tr1 switches on again, C1 discharges through R1 and Tr1, giving a shorter flyback. Tr2 acts to linearise the trace by Miller feedback.

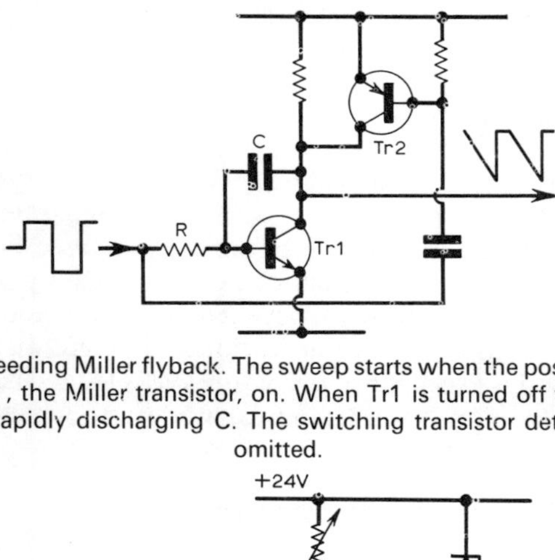

Fig. 7.17 Speeding Miller flyback. The sweep starts when the positive switching pulse turns Tr1, the Miller transistor, on. When Tr1 is turned off the same pulse turns Tr2 on, rapidly discharging C. The switching transistor details have been omitted.

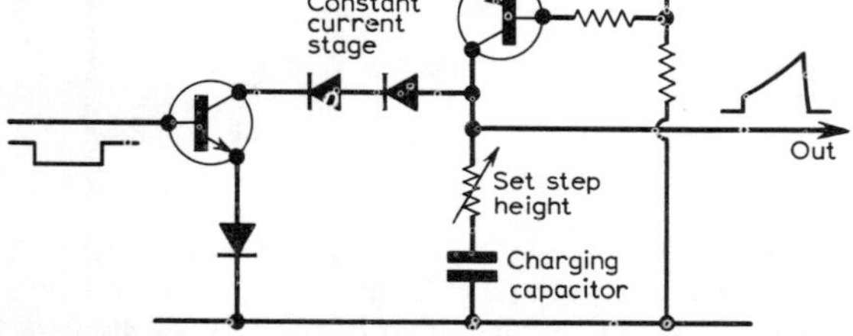

Fig. 7.18 Using a transistor as a constant-current supply—a simplified version of a timebase used in the Decca 'Transar' Marine Radar.

If there is no output, check that the squarewave generator is delivering a switching waveform to the timebase. Trace this waveform through to the switching stage, then check for waveform at the charging capacitor. If timebase is present at this stage, trace through amplifying stages to the output. If the time of sweep is low or high, but amplitude is normal, suspect the charging capacitor(s) and resistor(s) and any switch used to change sweep time or any potentiometer used to vary sweep time.

If the amplitude is too low, suspect that the squarewave is not wide enough for the sweep time so that the sweep has not time to rise far before flyback starts. The time constants of squarewave generator and sweep stage are varied together in oscilloscopes; the ganging arrangements may be faulty. If the sweep is non-linear, suspect the amplifier of a bootstrap stage, a leaky capacitor in any type of timebase or a low gain Miller amplifier.

Integrated Pulse Circuits

The use of integrated circuits in pulse work is now very considerable, and practically every function carried out by separate components can now be carried out more economically by ICs. The operations of shaping waveforms, for example, can be done by using CR networks in the feedback network between the output and the inverting input of an op-amp, and the integration and differentiating obtained by this method is excellent. If a squarewave is fed to such an op-amp integrator, the output is a triangular waveform which can be used as a timebase.

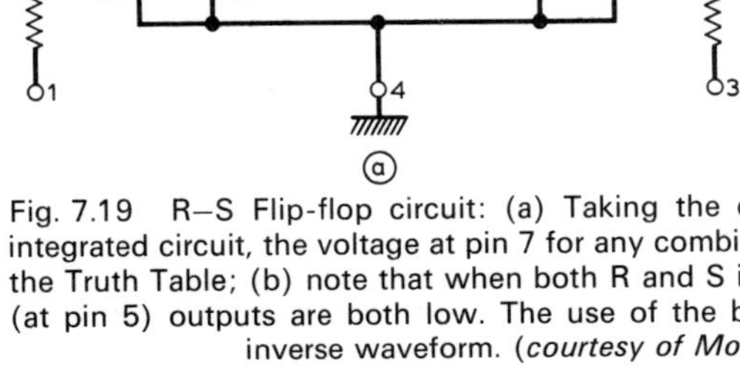

(a)

	R	S	Q^{n+1}
Pin	1	3	7
	Low	Low	Unchanged
	Low	High	High
	High	Low	Low
	High	High	Low

(b)

Fig. 7.19 R–S Flip-flop circuit: (a) Taking the output, Q, at pin 7 of this integrated circuit, the voltage at pin 7 for any combination of inputs is read from the Truth Table; (b) note that when both R and S inputs are high the Q and $\bar{Q}$ (at pin 5) outputs are both low. The use of the bar over a letter denotes an inverse waveform. (*courtesy of Motorola Ltd.*)

Most of the ICs available, however, are for pulse counting operations and computer logic applications. Schmitt triggers are available in IC form, but can be adapted from op-amps, as also can multivibrators of all types. The most important pulse circuit element found in IC form is the bistable, usually labelled as a flip-flop of some variety or other.

The RS flip-flop has a circuit typified by Fig. 7.19. With both inputs low (cutting off collector current), the output voltages at the common collector are unchanged; they remain as they were when last set or switched on. When one of the input terminals is switched on, the opposite pair of transistors switches off the transistor coupled to it, so increasing the total collector current by the cross-coupling resistor (*R*3 or *R*4) switching on the companion transistor.

The performance of the RS flip-flop may be summarised by a 'truth table', which shows the voltage level at one output for any combination of inputs. Note that when both inputs are switched high, both outputs are low.

The J-K Flip-flop

Computing requires that all counting circuits should complete their action on receiving a trigger or 'clock' pulse, and that the outputs should be decided by the combination of inputs, clock pulses and any reset signals used to change back the counters. Early flip-flop designs had the disadvantage that some combinations of signals could leave the output indeterminate, high or low according to the set pattern. To avoid this, the J-K flip-flop was evolved, a design too complex and expensive to use to any great extent when only separate components were used, but relatively easy and cheap to make as an IC.

The circuit of a J-K flip-flop and its truth table is shown in Fig. 7.20. It uses resistors and transistors as the circuit elements, and so belongs to the RTL (Resistor-Transistor Logic) family of circuits. Others are the DTL (Diode-Transistor Logic), TTL (Transistor-Transistor Logic) and ECL (Emitter-Coupled Logic), but the basic action of the J-K flip-flop is similar in any type of circuitry.

The circuit is composed of a flip-flop formed by *Q*1, *Q*2, *Q*3 and *Q*4 cross-coupled in the usual way. The clock input is applied through *Q*13 to the collectors of *Q*5 and *Q*6, whose emitters are connected to the flip-flop bases of *Q*2 and *Q*3 respectively. With a high voltage at the trigger input, the collectors of *Q*5 and *Q*6 are held at low voltage. Each trigger pulse is a negative pulse which allows the collectors of *Q*5 and *Q*6 to go high, and so allows *Q*5 or

*Q*6 to pass current to the base of *Q*2 or *Q*3, according to the voltage on the base of *Q*5 and *Q*6.

For example, if the SET input is low, Q output high, and CLEAR high, the collector-base junction of *Q*5 is forward biased (base positive to collector) and the stored charge in this junction will allow current to flow to the emitter junction whenever the collector of *Q*5 goes high, even if the base of *Q*5 is shut off at the same time. This happens whenever the negative trigger pulse cuts off *Q*13. *Q*5 conducts, pulling up the base of *Q*2, and making the flip-flop flop over to Q output low and Q high. By this time, the rise of voltage at the collector of *Q*13 has caused *Q*9 to conduct and hold the base of *Q*5 down. *Q*10, taking its base voltage from the collector of *Q*3, *Q*4 also holds down the base of *Q*5.

If both SET and CLEAR are high, *Q*5 and *Q*6 cannot operate, and there is no change in the $Q\bar{Q}$ outputs when the trigger pulse arrives. If SET and CLEAR are both low, then one of the two transistors *Q*5 or *Q*6 will have current in its base and the other will be held off by the action of *Q*10 or *Q*11. When the trigger arrives, the flip-flop will flop over making the outputs at Q and $\bar{Q}$ the opposite of what they previously were. With SET high and CLEAR low, the output at Q after the trigger is high; with SET low and CLEAR high the Q output after trigger is low.

The J-K flip-flop (so called because some manufacturers refer to the SET and CLEAR inputs as J and K) is used universally in computing, but in a large number of different forms. The type illustrated can be found with PRESET and PRECLEAR connections to the bases of *Q*1 and *Q*4, and these connections can be used to force the Q output to be high or low irrespective of other voltages applied.

Master Slave Flip-flop

The master-slave type of J-K flip-flop has an action in which the trigger causes one flip-flop to change state, and this in turn causes the change over of a second (slave) flip-flop. This type ensures complete isolation between the inputs and the outputs.

In use, J-K flip-flops can be synchronous or asynchronous. In synchronous operation, information is fed in to the SET and CLEAR inputs and a trigger is taken to each flip-flop. In the asynchronous method operation used for counting random pulses. SET and CLEAR terminals are earthed and the count voltage is taken to the trigger terminal. More details of counting circuits using J-K and other flip-flops are given in Chapter 8.

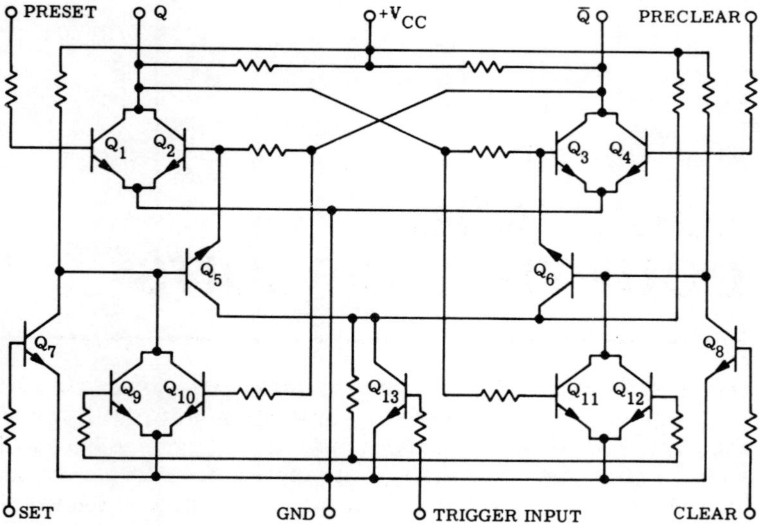

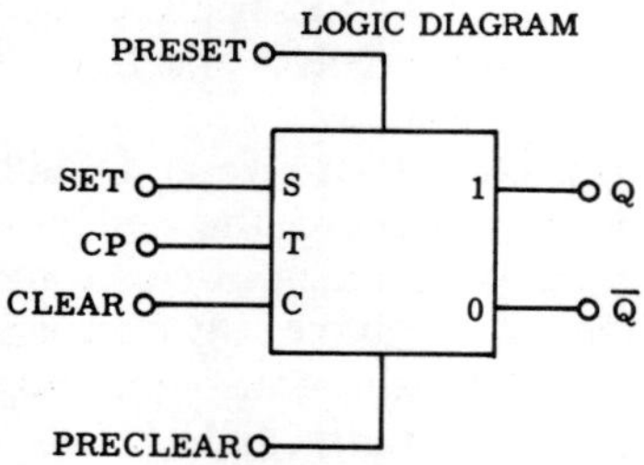

TRUTH TABLE

t_n		t_{n+1}
SET	CLEAR	Q
1	1	Qn
1	0	1
0	1	0
0	0	$\overline{Q}n$

HIGH VOLTAGE = 1
LOW VOLTAGE = 0

Fig. 7.20 J–K flip-flop: Circuit diagram, logic diagram and truth table. Note that this is only one possible way of obtaining J–K flip-flop action, others are possible. (*courtesy of Motorola Ltd.*)

CHAPTER EIGHT

LOGIC AND COUNTING CIRCUITS

THE LAST FEW YEARS have seen an enormous extension of the use of logic and counting circuits in all sorts of applications. Logic circuits are used wherever decisions of an automatic type have to be made. For example, the amount of petrol injected into a cylinder may depend on the position of the piston, the load on the wheels, the depression of the accelerator, the air temperature and humidity and other factors: a circuit which accepts signals sensing all of these factors and which then regulates the amount of petrol injected is a logic circuit.

Summing up, then, the logic circuit is one which gives a definite output for some arrangement of inputs chosen by the designer. Logic circuits are used in counting because so much of electronic counting uses a binary scale (scale of two) in which the only numbers are 1 and 0. This is particularly convenient, because transistors may then be operated bottomed or cut-off, with no intermediate states.

The logic functions of most importance are the OR, AND and NOT. The OR operation means that there is an output from the device if one input OR the other is on. The definition of the 'output' or the 'on' may be as a high voltage out or a low voltage out, this is a matter of definition. One way of writing the OR operation is 1 OR 1 $\rightarrow$ 1.

Another notation is Boolean Algebra, devised before electronics was known. In Boolean algebra, the action of the OR gate, as the OR device is called, is described by the equations: $0+0 = 0$; $0+1 = 1$; $1+1 = 1$, where $+$ means OR, and the equals sign means 'gives' rather than the more usual 'equals'.

The AND operation means that an output is 1 only when all of the inputs are 1; 1 and 1 and 1, etc, gives an output of 1. In Boolean this becomes $1.1 = 1$; $0.0 = 0$; $0.1 = 0$, where the dot indicates AND rather than the more usual multiplication. The NOT operation is simply inversion; a 0 input into a NOT gives a 1 output, and a 1 input gives a 0 output; this operation is written as a bar over the input being inverted, for example, $\overline{1} = 0$; $\overline{0} = 1$.

The NOT operation can be carried out at the same time as the other operations, giving us NOR (Not OR) and NAND (Not AND) gates whose internal operation is described in the companion to this volume, *Understanding Electronic Components*.

Logic Symbols and Action

Rather than draw out the complete circuit of a gate each time we use it, we use symbols for the action; some of these symbols are shown in Fig. 8.1. The main shape shows whether AND or OR operation is used, and the circle is used to indicate the negative output, making the gates NAND or NOR. We can now forget about the circuitry inside the gates, unless we are trying to use them in an unusual way, and concentrate on their action in circuits.

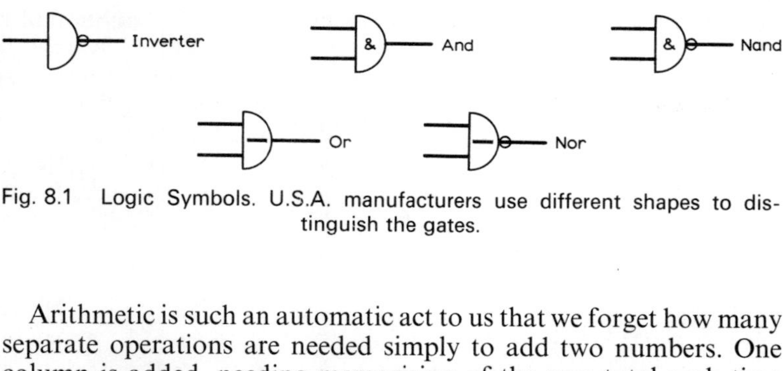

Fig. 8.1 Logic Symbols. U.S.A. manufacturers use different shapes to distinguish the gates.

Arithmetic is such an automatic act to us that we forget how many separate operations are needed simply to add two numbers. One column is added, needing memorising of the new total each time another number is added in. The last digit is written down, and the 'carry' memorised to be added to each number in the next column, and so on. The elements of this process are the addition, the temporary storage of the carry figure and the presentation of the output in some memorised form to be read out when convenient. In electronic counting, these operations are carried out by gates and flip-flops.

Binary Addition

In binary addition, all numbers into the system are the binary numbers 0 and 1. The addition is exactly as it is with decimal numbers except that the carry takes place when the number 2 would normally have been reached, so that $1+1 = 10$ (the + sign having the normal meaning of addition). Table 8.1 shows some examples of binary addition. Each digit, 0 or 1, is termed a bit (binary digit), and the circuits needed for carrying out the addition of bits are the half-adder and the full adder.

A half-adder adds two bits, one presented at each input; but has no input for adding in a carry, though a carry can be taken from the output. The truth table for a half-adder is shown in Fig. 8.2 along with a combination of two NOR gates and a NAND gate which gives the required output. Note that the inverse bits $\bar{A}$ and $\bar{B}$ are also needed to perform the half-adding.

The reader should check through the truth table inputs to see how the half-adder works; for example, a 1 at A and a 0 at B means a 0 at $\bar{A}$ and a 1 at $\bar{B}$. The output of the AB NOR gate is therefore 0 and the output of the $\bar{A}\bar{B}$ NOR gate is also 0. 0 appears at the carry output, and the two zeros appear at the input of the NAND gate, giving a 1 at the output as the result of the addition.

The half-adder would be found dealing with the first column of a

Decimal	Binary		Decimal	Binary
1	1		2	10
2	10		+5	101
3	11		——	——
4	100		7	111
5	101			
6	110		3	11
7	111		+8	1000
8	1000		——	——
9	1001		11	1011
10	1010			
16	10000		7	111
20	10100		+10	1010
32	100000		——	——
64	1000000		17	10001
100	1101000			
			7	111
			+3	11
			——	——
			10	1010

Table 8.1 Binary Scale and Binary Addition

A	B	S	C_{out}
0	0	0	0
0	1	1	0
1	0	1	0
1	1	0	1

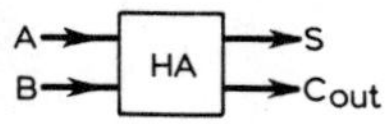

(a)

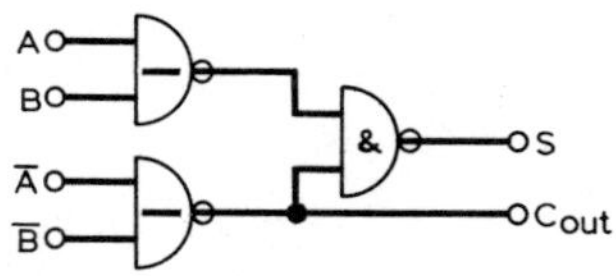

(b)

Fig. 8.2 Half adder: (a) logic diagram and truth table, (b) Half adder made from NOR and NAND gates.

binary adding system, but the remaining stages must all be able to accept three inputs, the third being the carry over input from the previous stage. This is done by the full adder, which can be built up in several different ways. Fig. 8.3 shows a method of making up a full adder from half-adders and an OR-gate. The inputs are the numbers and their inverses and the carry (with inverse), and the outputs are the sum, carry and carry inverse.

Parallel Adders

In a parallel adder, the binary numbers to be added are presented at the same time to the adding circuit which then adds them and produces an output which can be displayed or used in another stage. To each digit or bit of the number, one full adder will be required except for the final figure which requires only a half-adder, as explained earlier. When the numbers are applied to such an adder, adding takes place and the carry follows in each stage, so that the addition is not complete until the carry has taken place in each stage.

It would be possible for a carry to be generated in the first half-adder which would then cause a carry to take place in each of the other stages, a process known as 'rippling'; for this reason, an adder of this type is often known as a 'ripple-carry' adder. This ripple process determines the time which is required for an addition pro-

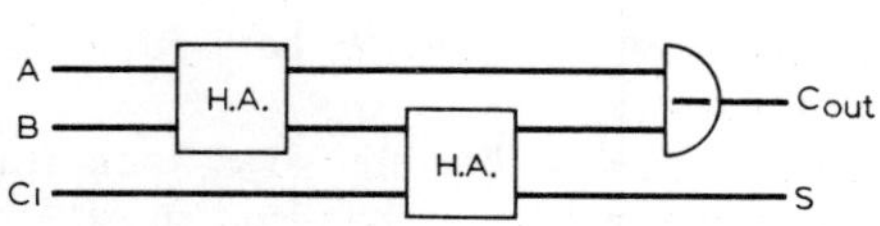

(a)

(b)

A	B	CI	S	C_{out}
0	0	0	0	0
0	0	1	1	0
0	1	0	1	0
0	1	1	0	1
1	0	0	1	0
1	0	1	0	1
1	1	0	0	1
1	1	1	1	1

Fig. 8.3 Making up a full-adder: (a) circuit using two half-adders and an OR-gate, (b) Truth table.

cess to be fully completed. Fig. 8.4 shows a simple form of ripple carry adder.

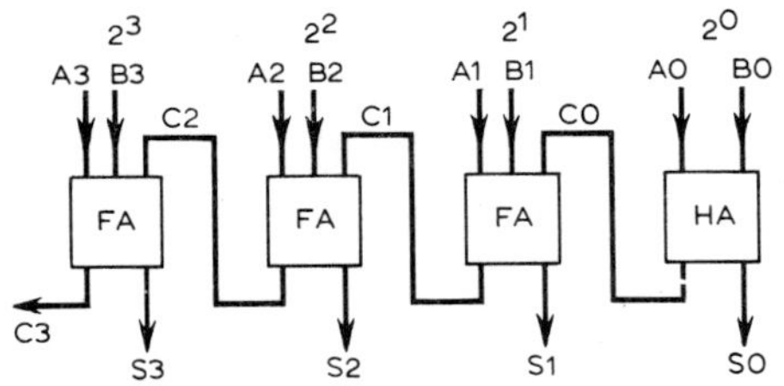

Example: 1001 + 0101 = 11110

A3	A2	A1	AO	B3	B2	B1	BO	S3	S2	S1	SO
1	O	O	1	O	1	O	1	1	1	1	O

CO = 1
C1 = O
C2 = O

Fig. 8.4 Ripple-carry (parallel) adder. The lowest numbers (least significant bits, or LSB) are loaded into the half-adder at the right hand side, and the highest (Most Significant Bit, or MSB) to the full adder at the left hand side, with intermediate bits loaded according to their significance (power of two). The number of stages determines the size of numbers which can be handled, but only two numbers can be added in one operation.

Serial Adders

A serial adder adds the digits of two numbers stage by stage in the way in which children are taught to add, by adding the lowest figures, memorising the carry, adding to the sum of the next higher digits and so on. This method of addition requires that each number to be added in must be stored (in a register, see later) so that it can be fed one digit at a time into the adder under the control of a trigger, or clock, pulse. A sum register is also required so that each digit of the sum can be stored for readout, and a temporary store is also needed for the carry which is generated in each addition and inserted into the next stage. Fig. 8.5 shows the form of this type of adder. Note that only one full adder circuit needs to be used, at the expense of providing registers for storage.

One of the registers can be eliminated if the sum is fed back into one of the registers which was used for storing the input numbers. The entry of numbers into the registers may be done in steps (serial entry) or into all stages simultaneously (parallel entry), though the adding will always be serial. Serial addition is slower but simpler than parallel addition, and is very widely used together with parallel entry to the registers.

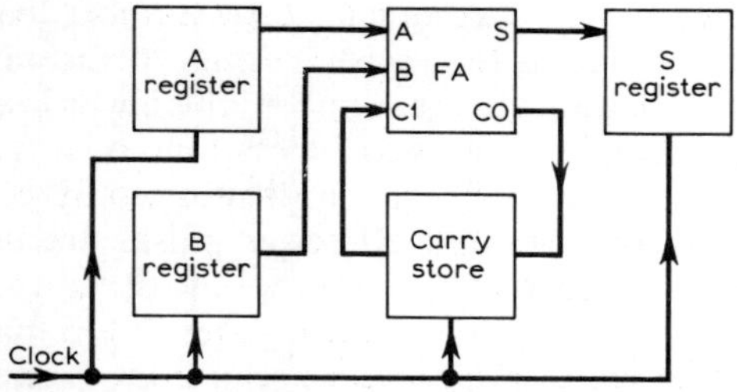

Fig. 8.5 Serial adder. Unlike the previous adder, this requires a clock (trigger) pulse, and an operation takes place each time the pulse arrives, usually on the trailing edge of the pulse. The carry store is used to take the carry digit from one operation and present it to the C1 input of the adder at the next clock pulse. The S-register can be eliminated by loading the output back into the A-register.

Shift Registers

A shift register is a set of flip-flops connected in series and used for the temporary storage of binary numbers, one flip-flop for each bit. If the stages of the register are *R*0, *R*1, *R*2 etc, up to the greatest number required, then we can use these labels to describe the size or 'significance' of the bit entered. The *R*0 register will contain the bits 0 or 1, representing the numbers 0 and 1; the *R*1 register contains bits 0 and 1 representing numbers 0 and 2, the *R*2 contains bits representing 0 or 4, and so on (Fig. 8.6)

If the registers are connected in series, with each Q output taken to an S input, and each $\bar{Q}$ output to the next C input, then the input is serial, and one bit loaded into the highest numbered register will be shifted one stage to the right on each clock pulse. For example, if we have shift registers 0–15 (a total of 16 registers) and a 1 is applied to register *R*15, then 16 clock pulses later that 1 will appear in the *R0* register, correctly showing the binary number 1. If a bit, 0 or 1, is loaded in at each one of the sixteen clock pulses, then the

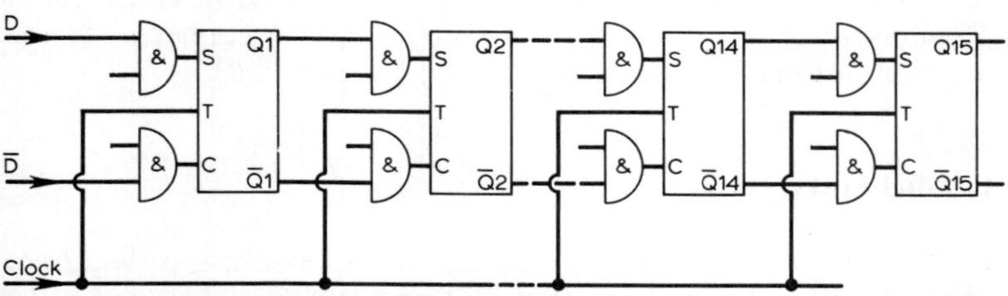

Fig. 8.6 Shift-right register. The AND-gates shown are not strictly necessary in this application, but are usually used in the more complex circuits, and often are built into the register.

register will end up with a complete 16-bit number loaded so that a readout of each stage of the register using, for example, lamps to indicate 0 or 1 will indicate the number which was loaded in.

This is an example of a parallel readout of a serially loaded register. Note very carefully that, in loading, only 16 clock pulses must be used in our example. If fewer pulses are used, then the number loaded is incorrect because the lowest value bits have not reached the appropriate parts of the register; if too many pulses are used the number is again incorrect because the higher value parts of the register have more zeros than they should.

This business of matching the number of the clock pulses to the register capacity is most important, and accounts for some of the procedures which puzzle trainee computer programmers regarding the statement of the length of entry of each piece of information. It is, however, possible to put the principle to good use to enable serial readout by feeding in another 16 clock pulses with no numbers presented at the inputs. As the clock pulses activate the flip-flops, the numbers will be unloaded in order with the least significant (0 or 1) first and the largest, 2^{16} (if there is a 1 in the 16th register) last.

The last digit, 1000000000000000 in binary, represents the decimal number 66,536, and the full register with 1-bits loaded all the way down represents the decimal number 131,071, so that a 16-bit register can be used for a fairly respectable number of computing applications. Note that the straightforward connection outlined above permits shifting in one direction only, to the right as conveniently drawn with *R*15 on the left and *R*0 on the right.

End-Around-Shift

It is sometimes desirable when serial readout is used *not* to remove the stored information. In this case, the output of the *R0* register is connected to the input of the *R*15 register during readout so that as each pulse is read out it is loaded in again by the same clock pulse. The pulses can be read through a buffer amplifier connected to the junction of the *Q* output of *R0* and the *S* input of *R*15.

Parallel Entry

Parallel entry into a J-K flip-flop has to be done using the direct connections to the flip-flop (PRESET and PRECLEAR) and must also be done at a time when the clock voltage is steady, not during a pulse. To ensure this, the parallel entry has to be achieved through gates which are opened by pulses (enabling pulses) only during a period of steady clock voltage.

It is possible to switch one clock pulse from the clock line to the enabling line when entry has to be made, otherwise enabling pulses can be obtained from the clock pulses through a time delay circuit. For gating, two NOR gates are required for each register, as shown in Fig. 8.7

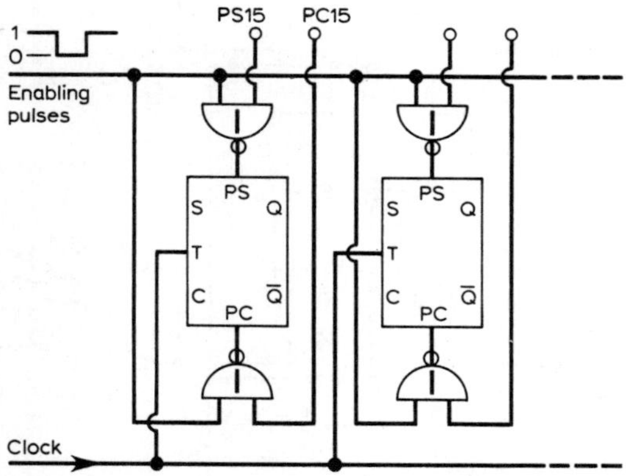

Fig. 8.7 Parallel entry into shift register. Only the parallel entry connections are shown here to simplify the diagram.

Left-Right Shift

Right shift in a serial input register is easily achieved with no gating. In many applications, shifting to the left is needed; for example, multiplication is carried out by adding and left-shifting operations. One way of achieving left-shift is shown in Fig. 8.8, where the interconnections are partly shown by labelling to avoid having too many circuit lines. Information can be fed in at *DR* for shifting right or at *DL* for shifting left, and the voltages on the SHIFT LEFT or SHIFT RIGHT lines decide in which way the shift shall take place when the trigger occurs.

In normal operation, the shift pulses are complementary (if $L = 1$, $R = 0$, and if $L = 0$, $R = 1$), but it is also possible to use the register with both *R* and *L* low. In this state, each clock pulse toggles the flip-flops, meaning that each bit stored is inverted. If a 1 is stored in any part of the register, it will become a 0 after toggling, and each 0 will, similarly, become a 1. This turns the number stored

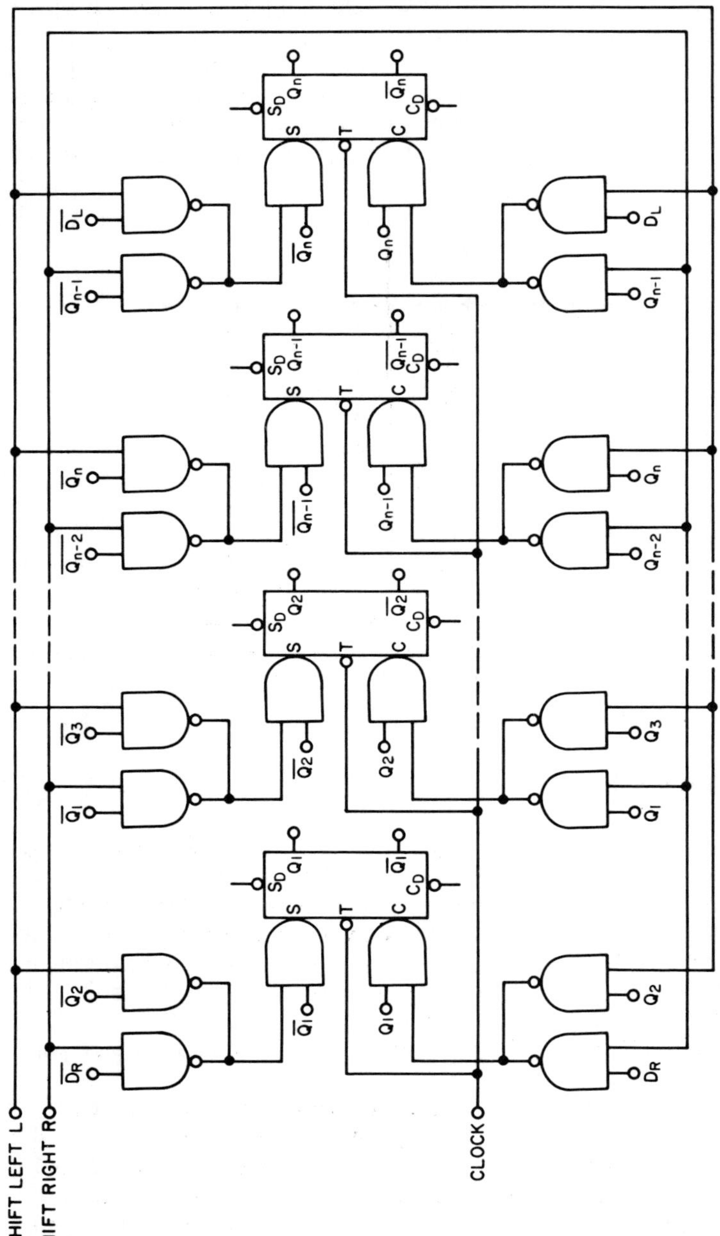

Fig. 8.8 Shift-right-shift-left gating. Note that the precise gating used varies with the type of IC logic system (DTL, RTL, ECL etc.) (*courtesy of Motorola Ltd.*)

into its inverse, and is useful for subtraction, because in some codes adding the inverse of a number is equivalent to subtracting the number itself.

For example, the subtraction: $25-5 = 20$ can also be done by taking $25+95 = 120$ and discarding the 100 term. The operation here has been to take the complement of 5 with a power of ten greater than the largest number handled (25 in this case), so giving us $100-5 = 95$. On adding, the 100 which has been introduced can then be discarded. This may seem cumbersome, but it saves having to have a completely different form of circuit for subtraction.

In binary, the procedure is more complicated: 25 becomes 11001 and 5 is 101; the complement of 101 is 1 + the inverse, giving 011 (010+1), and the figure has then to be made up to the same number of digits as in 11001, giving us 11011, since the making up is with 1-bits. Addition now gives us 110100, and the highest digit is ignored, giving 10100 (decimal 20).

The complications arise because the normal binary code is not 'self-complementary', meaning that the simple inversion of 0's and 1's does not give a true complement. For decade counters (working in tens) several self-complementing codes using the binary digits 1 and 0 have been devised, and are used in small calculators with normal binery reserved for the larger machines.

Decade Counters

For many purposes apart from large computors, a decade (scale of ten) counter is useful. Measurements of frequency, for example, can be made by operating a gate for a fixed time, accurately controlled, and counting the received pulses in a decade scaler. Voltage and current values can be converted to a series of pulses, the number being proportional to the size of the voltage or current, and counted to give a digital readout of voltage or current.

A large variety of counting circuits can be devised to give a count of ten, and a large number of codings has also been devised, each with its own particular merits and disadvantages.

Decade Ripple Counters

In a ripple counter, the output of one flip-flop drives the output of the next. This method uses fewer gates than other counters, and also has the advantage for high frequency operation (more than 15MHz) that only the first flip-flop is operating at the fastest rate. The output

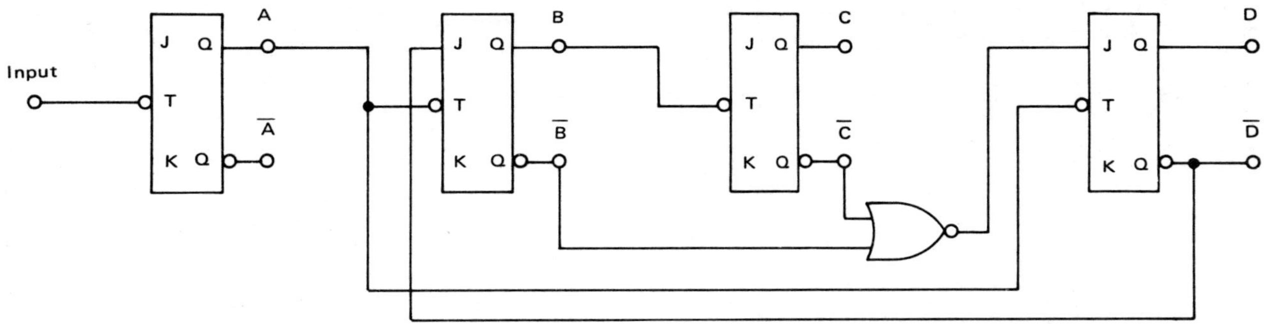

COUNTER SEQUENCE

	A	B	C	D
0	0	0	0	0
1	1	0	0	0
2	0	1	0	0
3	1	1	0	0
4	0	0	1	0
5	1	0	1	0
6	0	1	1	0
7	1	1	1	0
8	0	0	0	1
9	1	0	0	1

UNUSED STATE ANALYSIS

Unused States					Next State			
A	B	C	D		A	B	C	D
0	1	0	1	→	1	1	0	1
1	1	0	1	→	0	0	1	0
0	0	1	1	→	1	0	1	1
1	0	1	1	→	0	0	1	0
0	1	1	1	→	1	1	1	1
1	1	1	1	→	0	0	0	0

DECODING LOGIC

0	$\bar{A}\ \bar{B}\ \bar{C}\ \bar{D}$
1	$A\ \bar{B}\ \bar{C}\ \bar{D}$
2	$\bar{A}\ B\ \bar{C}$
3	$A\ B\ \bar{C}$
4	$\bar{A}\ \bar{B}\ C$
5	$A\ \bar{B}\ C$
6	$\bar{A}\ B\ C$
7	$A\ B\ C$
8	$\bar{A}\ D$
9	$A\ D$

Fig. 8.9 Decimal ripple counter. (*Motorola Ltd.*)

is not available until all of the counts have taken place, but this is not necessarily a disadvantage. Three flip-flops can perform a division by eight, and four a division by sixteen, so that a decade counter must use four stages and amend the wiring so that the counter returns to its original state at the count of ten.

The ripple counter of Fig. 8.9 uses binary code to give a decimal count by an ingenious interconnection of the counters with one NOR gate, so that several possible counter states are unused. In the figure, the output of *A* changes on each received trigger pulse, and triggers *B*. The same trigger is also taken to *D*, but the 1 inputs from $\bar{B}$ and $\bar{C}$ (because *B* and *C* are at 0) through the NOR gate place a 0 on the *J* terminal of *D* and prevent the trigger changing it over in the early stages of the count. Not until the normal output of *A*, *B*, *C* each give 1 can the next trigger affect *D* (at the transition from 7 to 8), which then changes over to give a 1 at *D* and 0 at $\bar{D}$, bringing down the *J* input of flip-flop *B*.

On the next trigger in (count of 9), *A* goes high and the next input causes both *A* and *D* to change over, *B* being held at 0 by the voltage on its *J* connection; *C* is already at 0. The last pulse (10) therefore returns all of the flip-flops to zero, ready for another decade count, and the trigger from *D* can be passed to the next decade. Binary-coded-decimal (BCD) counters of this type can now be obtained in completely integrated form, often with outputs from intermediate stages so that counts of 2 and 5 can also be used. Scale of 12 and 16 counters can also be obtained.

Decoder Stage

Decade counters, whatever their type, require a decoder stage before the state of the count can be displayed in a way which can be read immediately, such as on a gas-display tube. Each code used needs a different decoding logic and it is necessary to buy matching counters and decoders unless the code and drive conditions of units of different manufacture match up well. Most decade counters use BCD, and BCD decoders are readily available and reasonably well standardised.

The decoders consist of inverters and NAND gates arranged so that the states of the decimal counter give the decimal count, at the pins numbered 0 to 9, in the form of voltages present or absent. In the BCD-to-decimal decoder shown in Fig. 8.10, the decimal count is given by the appropriate leadout switching from 1 to 0, and the whole set of eight inverters and ten NAND gates is integrated into a single package.

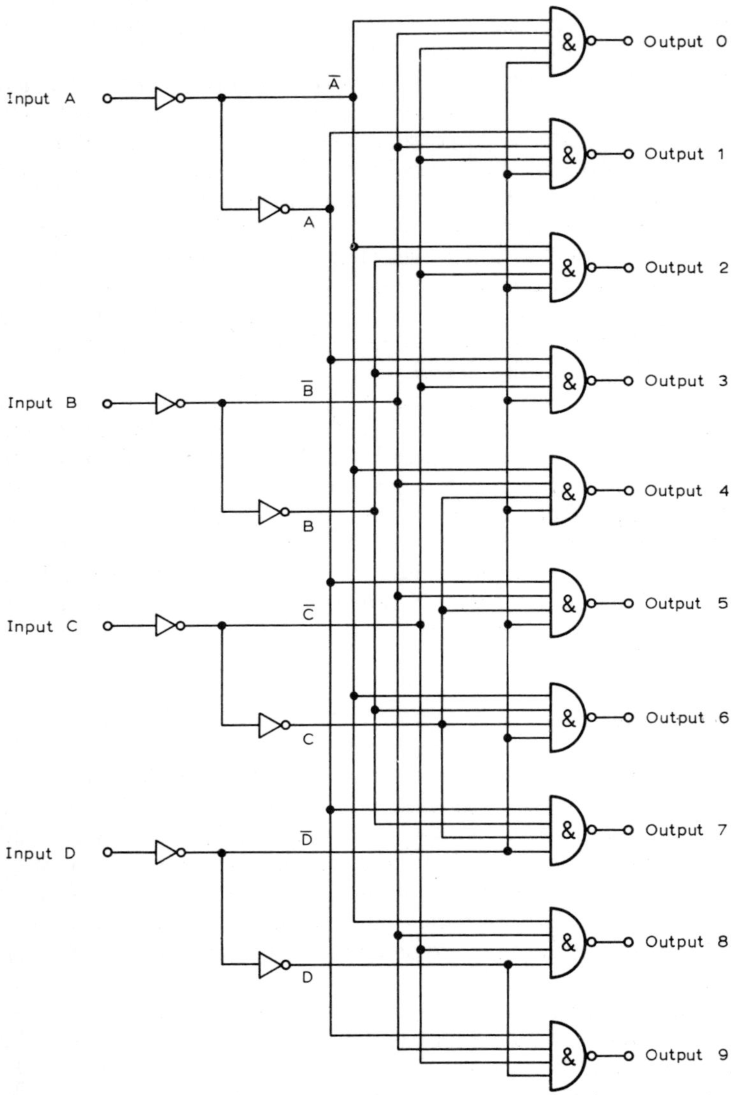

Fig. 8.10 FJH 261 BCD-to-decimal decoder. (*courtesy of Mullard Ltd.*)

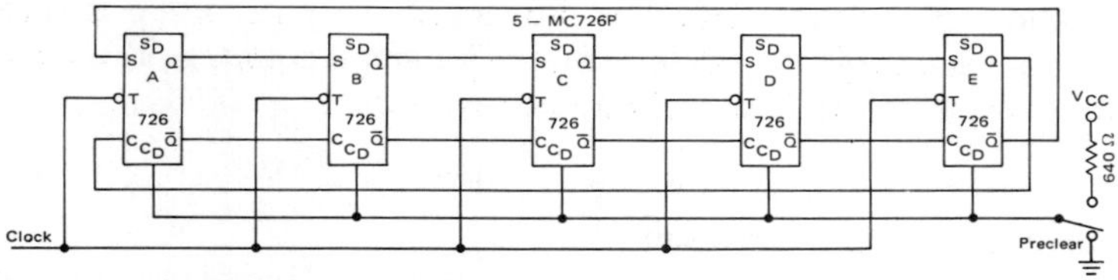

Truth Table

Count	E	D	C	B	A
0	0	0	0	0	0
1	0	0	0	0	1
2	0	0	0	1	1
3	0	0	1	1	1
4	0	1	1	1	1
5	1	1	1	1	1
6	1	1	1	1	0
7	1	1	1	0	0
8	1	1	0	0	0
9	1	0	0	0	0

Output Decoding

Count	Logic
0	$\bar{E}\,\bar{A}$
1	$\bar{B}\,A$
2	$\bar{C}\,B$
3	$\bar{D}\,C$
4	$\bar{E}\,D$
5	$E\,A$
6	$B\,\bar{A}$
7	$C\,\bar{B}$
8	$D\,\bar{C}$
9	$E\,\bar{D}$

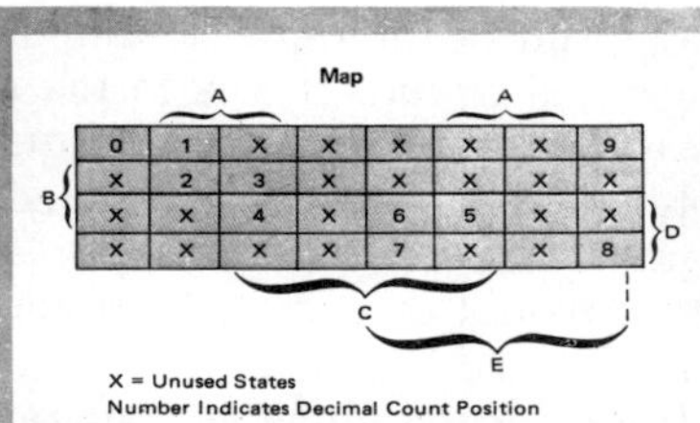

Fig. 8.11 Decade shift counter using RTL circuitry. (*Motorola Ltd.*)

Synchronous Counters

Because of the vast variety of decade counters, it is not possible here to give a full account of even a representative sample of types, but two examples of synchronous counting must be shown. A synchronous (clocked) counter has the advantage over the ripple type that all the changes take place when a trigger (clock) pulse is applied. This requires, of course, that the clock can cope with driving all the stages.

The decade shift counter of Fig. 8.11 is a clocked counter of a type whose outputs are particularly easy to decode (especially to an analogue signal, see later) and which requires no gating between stages. The feedback from counter *D* to the input of *A* is to make the counter self-correcting: if an accidental misfiring causes the flip-flops to assume a state for which there is no code (an unused state) then the circuit round the feedback loop causes an eventual return to normal counting, so avoiding the necessity of clearing and starting again.

The decade shift counter uses five flip-flops; the ring counter uses as many flip-flops as the scale of counting, so that for a scale of ten, ten flip-flops are needed. The ring counter is particularly useful for odd scales, but the straightforward ring counter, where each Q and $\bar{Q}$ is connected to the next J and K respectively, is not self-correcting. A self-correcting version of the decade ring counter is shown in Fig.

8.12, and the counting sequence shows how easy decoding is; the output of each stage is a decimal digit. For small counters, this may be more economical than BCD and converters.

Analogue to Digital Converters

One result of the rapid development in digital circuitry caused by integrated circuits has been the application of digital methods in measurement and control. If quantities such as voltage, weight, volume, rate-of-flow, speed, acceleration, etc., can be coded in digital form, they can then be used in digital logic circuits and counters and used directly in computing with all the flexibility of digital methods.

In some cases, analogue computing, involving the differentiation and integration of the waveforms themselves may be preferable, but digital computing has become more common since integrated circuitry made digital methods competitive in price, and analogue computing will not receive further attention in this book.

Digital methods have the advantage that the maximum error in counting will not be more than one digit, and that the display is very much easier to read. Quantities such as frequency are, of course, already in a form suitable for counting, but some means of limiting the count is needed. The basic methods of digital frequency measurement involve either counting the number of complete cycles of the unknown frequency which occur within a standard time interval or counting the number of cycles of standard frequency which occur within the time of a cycle of the unknown frequency.

In each case, one waveform of comparatively low frequency is being squared and used to gate a number of cycles of a higher frequency waveform. Frequencies which are too high to be dealt with by counting directly can be mixed (heterodyned) with a known frequency, to produce a different frequency which can be more readily counted; the original frequency can then be found by calculation.

Voltage, however, is not naturally 'countable', and most physical quantities are most easily measured in the form of voltage. Analogue-to-digital converters are used to change a voltage-level signal into a stream of pulses whose number is exactly proportional to the voltage.

One quantity which can be more easily dealt with directly is the rotation of a shaft. On the simplest possible level, a cam on a shaft operating a microswitch will produce a pulse on each revolution. For systems where the position of the shaft may have to be known to

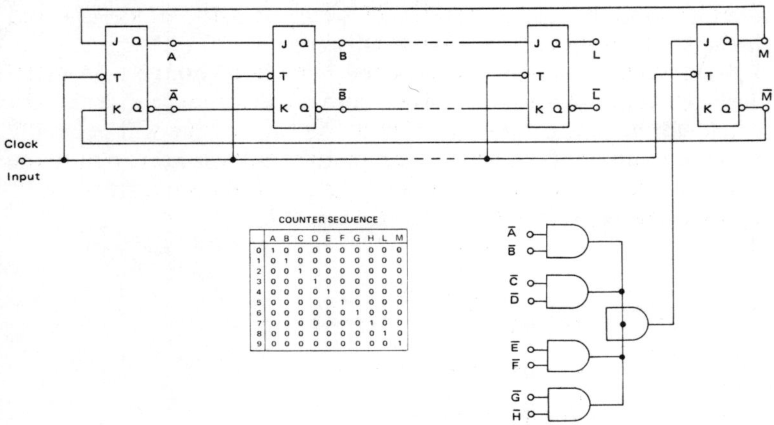

COUNTER SEQUENCE

	A	B	C	D	E	F	G	H	L	M
0	1	0	0	0	0	0	0	0	0	0
1	0	1	0	0	0	0	0	0	0	0
2	0	0	1	0	0	0	0	0	0	0
3	0	0	0	1	0	0	0	0	0	0
4	0	0	0	0	1	0	0	0	0	0
5	0	0	0	0	0	1	0	0	0	0
6	0	0	0	0	0	0	1	0	0	0
7	0	0	0	0	0	0	0	1	0	0
8	0	0	0	0	0	0	0	0	1	0
9	0	0	0	0	0	0	0	0	0	1

Fig. 8.12 Decade ring counter, DTL circuitry. (*Motorola Ltd.*)

a fraction of a degree, the shaft can carry a disc which has a pattern of holes in which the outermost holes represents the lowest digits of a binary number and the innermost holes the highest digits.

The number of 'rings' of holes needed depends on the accuracy required; if six rings are used, the innermost consisting of one hole and one solid, then a six-bit number can be coded, and the highest six-bit number is 111111, or 127 in decimal. In such a case, the position of the shaft could be coded to an accuracy of 365/127 degrees. The accuracy of such a device depends on the precision with which the holes can be formed and read, the shape of the hole is that of a rectangular slot, each hole being the same size as the solid portion which represents the 0.

One good method is to use a clear glass or plastics disc and print a pattern of bars and spaces on to it photographically. The readout is usually by a light source shining through the clear or cut-out portions and a phototransistor detecting the light pulses and so converting them into voltage or current pulses to be counted.

Voltage Coding

Voltage coding has been carried out with a very wide variety of circuits, and only an outline can be given here. In simple cases, where only a few digits are needed to represent voltage levels which do not have to be controlled or monitored exactly, a set of biased diodes can be used as the coder. Each diode is biased to a different reference voltage, and a resistor in series with the diode is wired through an amplifier to a Schmitt trigger.

If the voltage applied exceeds the reference voltage to a given diode, current flows in the resistor, and a trigger pulse is generated which can be entered into a counter. As the voltage will generally trigger a number of diodes at once, parallel entry is needed, and the code used here is the 'ring-counter' type, with the number of digits representing the size of the voltage. (Fig. 8.13)

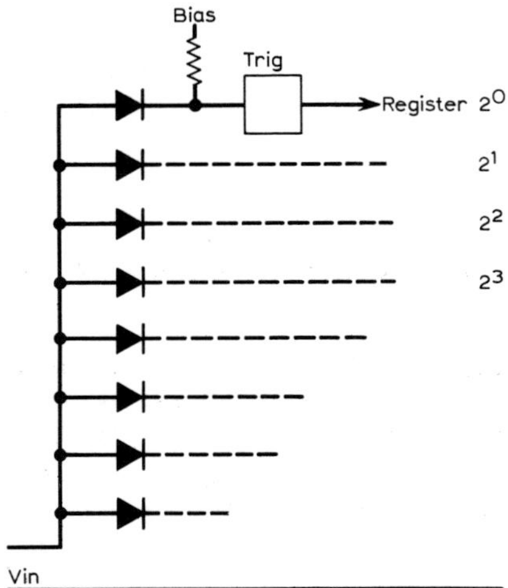

Fig. 8.13 Coding a voltage level by biased diodes.

Serial methods are more popular, and the timebase encoder is a good example. A simple type uses a starting pulse to gate a set of clock pulses and also to form (by integration) a linear ramp of voltage. The ramp is applied to a comparator, usually an op-amp, together with the voltage to be measured and, when the voltages are equal, the comparator shuts off the gate. The number of pulses which has passed through is then proportional to the voltage, since a higher voltage will take longer to reach and more pulses will pass the gate. This technique is quite adequate where great accuracy is not required; the limitation is the linearity of the ramp voltage.

For more accurate *A* to *D* conversion, the timebase may be generated by the digital pulses themselves. The digital pulses are integrated, using a long time constant, until the total voltage generated equals the applied voltage and the comparator then detects the identical levels and shuts off the pulses. Alternatively, the digital pulses may be used to switch pulses into the comparator, each pulse

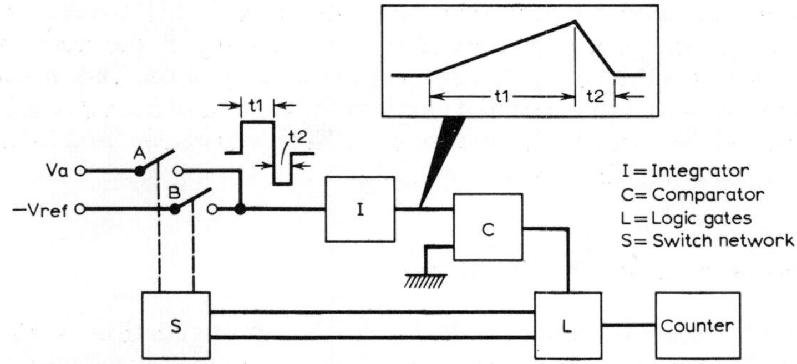

Fig. 8.14 Dual slope converter. To start the circuit, switch A is activated for a time t1. After time t1; the switching network S changes over, closing B and opening A, so that the integrator feeds a falling voltage to the comparator C. When this reaches zero, the logic stage L activates the switches again and opens B. During these processes, clock pulses are fed out.

being of slightly higher voltage, until equality is again reached. Note that this type of converter is actually using *D* to *A* conversion (pulses into voltage level) in a feedback type of circuit.

Another method of integration, which can also be used for *D* to *A* conversion, uses each pulse to add a known amount of charge to a capacitor. As the voltage across the capacitor rises linearly with the charge placed on it, this voltage can be applied to the comparator in the usual way. Also using the idea of capacitor charge and discharge is the dual-slope converter. This uses an integrator circuit based on an op-amp whose output is a linear rise or fall of voltage when a step of voltage is applied at the input.

The integrator is arranged so that the input may be either the unknown voltage or a fixed reference voltage, and the switching is arranged to occur at a definite time, t_1, after the unknown voltage is applied. When the unknown voltage is applied at the input (Fig. 8.14), the step of voltage at the input causes a linear slope to be generated at the output.

After the fixed time, t_1, the input of the integrator is switched to the (negative) reference voltage, and the time taken for the voltage to return to zero is measured, t_2. The ratio t_1/t_2 is equal to the ratio $V_{\text{reference}}/V_{\text{unknown}}$, and this is true whatever time constants are used in the integrator. As it stands, this measurement depends for its accuracy on the standards of time used, but this can be simply overcome by making *both* time measurements depend on one frequency standard.

Since it is the *ratio* of times which is important, any error in the frequency standard is cancelled out, with the result that the accuracy of conversion does not depend on the accuracy of the frequency standard but only on the consistency of the integration. This method lends itself well to integrated circuitry and can be carried out at low cost; for this reason, the dual-slope technique is widely used in low-cost digital reading instruments.

Successive Approximation Method

Another scheme, known as the 'successive approximation' method, has only become widely used now that ICs are readily available. It consists of comparing the unknown voltage with a series of accurately known voltages, and counting the number of steps needed to obtain a fit. At first the unknown is compared with $\frac{1}{2}V_{ref}$, where V_{ref} is the internal reference voltage.

If the unknown is greater than this value, it is then compared with the greater voltage: $\frac{1}{2}V_{ref}+\frac{1}{4}V_{ref}$; if it is less than $\frac{1}{2}V_{ref}$, it is next compared with $\frac{1}{2}V_{ref}-\frac{1}{4}V_{ref}$. The next comparison would be with $(\frac{1}{2}+\frac{1}{4}+\frac{1}{8})V_{ref}$ or $(\frac{1}{2}+\frac{1}{4}-\frac{1}{8})V_{ref}$ if the previous comparison had indicated that the unknown was greater than $(\frac{1}{2}+\frac{1}{4})V_{ref}$; or with $(\frac{1}{2}-\frac{1}{4}-\frac{1}{8})V_{ref}$ and $(\frac{1}{2}-\frac{1}{4}+\frac{1}{8})V_{ref}$ if the previous comparison had indicated a smaller value. This process can be continued, and each stage of comparison will add one bit to the output signal.

The accuracy obtained depends on the number of bits used and how closely the voltages can be compared. If there are n bits of code, then the value of the unknown can be found with an error of only $V_{ref}/2n$. For example, the converter described by Texas Instruments in Application Report B20/31 uses an 8-bit code with a reference voltage of 4V to give an accuracy of $\pm 4/256$V, which is about 15·6mV.

When *A* to *D* converters are used in measuring instruments, some provision is usually made for repeating the measurements at intervals; not too widely spaced (to avoid missing variations) nor so closely spaced that the last digit of the display is constantly flickering.

Digital to Analogue Conversion

D to *A* conversion is seldom needed to the same extent as *A* to *D*, but is used in several *A* to *D* convertors as part of a feedback network to keep the operation of the *A* to *D* converter linear. The methods of *D* to *A* conversion used depend very much on the coding used for digital information, but a few main types are worth noting.

The simple integrating circuit can be used where the digital code consists simply of a train of pulses (*unlike* the successive approximation methods above), and a development of this is the cup-and-bucket counter. In the circuit of 8.15, the pulse applied to *C*1 causes the rectifier *D*1, to conduct, so that *C*1 charges up. After the pulse, with the rectifier non-conducting, the charge re-distributes between *C*1 and *C*2, and the voltage across the plates of *C*2, is much less than the pulse peak voltage because the charge which was originally on *C*1 is now spread over a much larger capacitance.

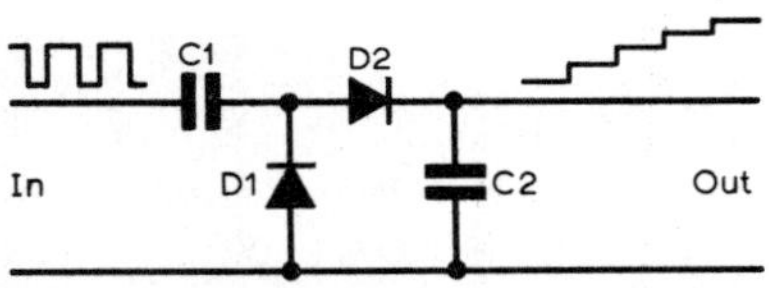

Fig. 8.15 Cup-and-bucket counter. C1 is the 'cup', C2 the bucket. D2 acts as a one-way valve preventing charge from moving back from C2 to C1 during the pulse.

At the next pulse, this process is repeated, but the amount of charge deposited is less, since the voltage difference between the pulse and the capacitor voltage is now less. The output of this circuit is a staircase waveform with steps of decreasing height; this can then be applied to a comparator circuit. In its basic form, the cup-and-bucket is a useful, simple, counting or frequency dividing circuit, with the counting ratio determined by the setting of the comparator and the ratio of capacitances.

It is non-linear, however, and cannot be used for *D* to *A* conversion unless the steps are made equal. This can be done by bootstrapping; for example, the voltage at the output can be used to make the pulse height greater at every pulse by the amount that the output voltage has risen, so that the amount of charge placed on *C*1 is the same at every pulse.

For the more usual digital codes, a ladder network converter is more common. The basic circuit is shown in Fig. 8.16. A network of resistors is used, with one end earthed, and the others capable of being switched between a reference voltage and earth. When a resistor near the bottom of the ladder is switched to V_{ref}, the effect is to raise the output voltage only very slightly. When a resistor near the top of the ladder is switched to V_{ref}, the effect is much larger, so that the ladder can be used to convert digital to analogue if the different resistors are switched by different digits, the lower value digits switching resistors at the bottom of the ladder and the high value digits switching the resistors at the top of the ladder.

Obviously there must be as many switched stages as there are bits in the code. Fig. 8.16 shows examples of two-bit numbers with the outputs available. The switches used in a practical example are solidstate devices, and Texas application Report B20/31 describes a converter using FET switching which uses a 9-bit code and gives an output within 1mV of the required value after 6μS of clocking the input.

Memory Circuits

Large computers use elaborate memory circuits which are the means of storing large amounts of binary information in a small space and without the need for active devices such as flip-flops. In such computers, information may be required from any part of the memory at any time (random access) and some means of coding the position of the information inside the memory is required. Such a code is called the 'address' of the information.

Where the memory consists of tape, the address of a piece of information is composed of the distance of the start of the information from the start of the tape, and the number of bits of information to be read. On a magnetic disc memory, the coding can consist of angular and radial position of a pickup arm. Such memories are random access, since any piece of information can be retrieved without having to read through all of the information; they can be 'written', 'read', and 'erased' with no effect on the rest of the stored information.

Magnetic core memories are also extensively used, and consist of grids of wire with a lump of magnetic material at each wire intersection. The wires are insulated from each other, and may be wound through or round the magnetic material to ensure closer coupling. Magnetisation or demagnetisation of the magnetic material takes place only when the current direction in each of the two wires through a core is correct, so that the address of a piece of information consists of directing current to the two wires concerned.

The information is destroyed when it is read out, but the circuitry can be arranged so that the information state (magnetised or unmagnetised) of each core is restored after 'interrogation'. The wiring problem in a large core memory is a formidable problem, since two wires are needed to address each bit.

Small counters and computers generally use the shift registers as the memory circuits, but a form of memory known as the 'read-

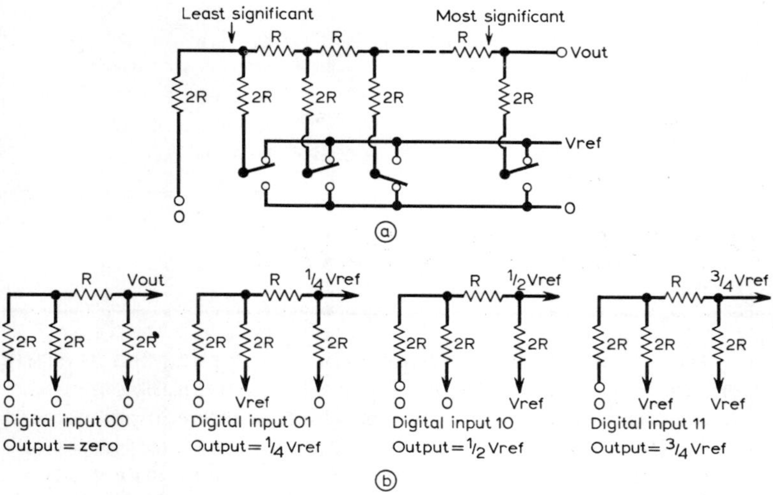

Fig. 8.16 Ladder-network converters. (a) general form, (b) illustrating action for a two-bit converter. (*courtesy Texas Instruments Ltd.*)

only-memory' (ROM) is also used. A read-only-memory has no method of storing information other than the internal connections made when it was manufactured, but the contents of the memory can be read out by address pulses whenever they are required.

At first sight, this might seem rather pointless, but the idea of having a small unit which will give a fixed response to a given set of inputs is extremely useful. A ROM can, for example, be used for any form of decoding, as it can be programmed to accept, for example, BCD and give any scale of output required. Another use is the generation of outputs which might be very difficult to generate by any other means.

For example, the readout of a counter in the form of letter shapes on a cathode-ray tube requires circuits capable of converting from number (usually BCD) information into waveforms which will produce the needed deflection and bright-up of the beam of the c.r.t. This can be carried out by the ROM, as can also the generation of other waveforms to the order of a digital code.

CHAPTER NINE

POWER SUPPLIES

ALL ELECTRONIC EQUIPMENT requires power supplies, and in nearly every case the power must be delivered to the equipment as d.c. Where current drain is low, and especially where equipment must be portable, batteries may be used; but in many cases, especially where valves are used, the source of power is the mains a.c. supply at voltages of 110V to 240V, and frequencies of 60 or 50 Hz depending on locality. In addition to these supplies there are such devices as fuel cells, solar cells, and the nuclear batteries for equipment which must work unattended for long periods.

Batteries

Batteries are classed as primary, which are used once and then discarded; or secondary which can be recharged. Each type relies on the conversion of chemical energy into electrical energy, the difference being that the chemical reaction in the secondary cell can more easily be reversed. Primary batteries have an impedance which becomes higher towards the end of life; secondary batteries have a very low impedance which varies little during discharge. A battery is formed by connecting unit cells; primary cells have impedances of 0·5Ω upwards, depending on size, secondary cells have impedances of 0·1Ω downwards.

For the more specialised applications, the type of primary cell used in the domestic torch or transistor radio battery, and the type of secondary cell used in the car battery are both ruled out; the first for its short life and rather high impedance, the second for its great weight and bulk.

Manganese batteries are composed of primary cells which have a lower impedance and a very much higher capacity for converting chemical energy. The capacity of a battery is usually stated in milliamp-hours (mAH); the best working current in milliamps multiplied by the time in hours for which that current can be sustained. This does not mean that any combination of milliamps and hours which gives the same figure is equally possible; a battery of 1,000mAH which can give 10mA for 100H does not give 100mA for 10H and certainly not 1A for 1H. The capacity of a manganese cell can be ten times that of a normal (Leclanché) cell of the same size, and the variation of voltage and impedance with life is much less. Naturally, the price is greater. Each single cell has a no-load voltage output of 1·5V.

Mercury cells can be made even more compact and are used in subminiature equipment. A mercury cell of 100mAH capacity can be made to a size of 15mm (about 0·6″) diameter by 15mm high, each cell having a no-load voltage of 1·4V.

Nickel-Cadmium cells are secondary cells, very much more compact than the old and familiar lead-acid accumulator, although not so compact in size for equal capacities as the manganese or mercury cells. A 1,000mAH nickel-cadmium cell, working at 1·24V no-load output would measure about 50mm (2″) diameter by 10mm (about 0·4″) high. An important point to note about nickel-cadmium cells is that the recharging current must on no account exceed the value laid down by the manufacturer; as the cells are sealed there is a risk of explosion due to pressure of accumulated gas if the charging current is exceeded.

Power Packs

A circuit designed to feed power to electronic equipment is known as a power pack. Most power packs supply both a.c. and d.c.; a.c. in valve circuits mainly for valve heaters, in transistor circuits for indicator lights and other auxiliaries, d.c. for the main supplies to both valves and transistors. There are two main steps in supplying these voltages—the supply of a.c. at the required voltage levels, usually by a transformer, and the conversion by the processes of rectification and smoothing of this a.c. into d.c.

The simplest power packs use the a.c. mains as the source of power; a typical circuit of a power pack is shown in Fig. 9.1 together with the typical performance. The mains voltage is stepped up by the transformer in two windings, each giving 300V r.m.s. and connected in series. This output winding is usually represented by the centre-tapped winding indicated.

The rectifier conducts when either anode is positive, and so the voltage output of the rectifier is as shown. The smoothing (integrating) circuit is a low-pass filter of two capacitors and an inductor. A

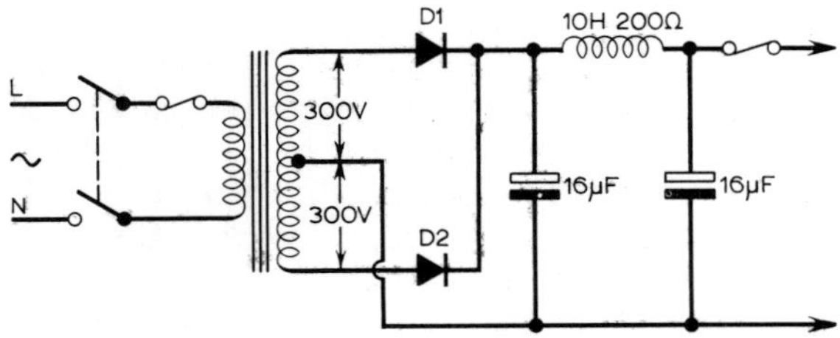

Current out	Voltage out
0	420V
100mA	370V
200mA	300V

Fig. 9.1 Simple rectifier circuit. The peak inverse voltage across each rectifier diode in this case is 840V, and the average current through each is half of the average output current.

circuit of this type is typical of millions of older radio sets, amplifiers and tape recorders; it is seldom found in so straightforward a form in more specialised electronic equipment.

Performance of a Power Pack

The figures most frequently used to measure the performance of a power pack are:

(1) the off-load voltage, which is the steady voltage measured with a negligible load applied,

(2) the maximum continuous current which can be drawn without causing overheating of the transformer or rectifiers,

(3) the regulation, which is the variation of voltage output with current. This may be quoted as the 'internal impedance' of the power supply, as if the power pack consisted of a perfect generator with a resistor in series (Fig. 9.2). The lower the internal impedance, the better the regulation,

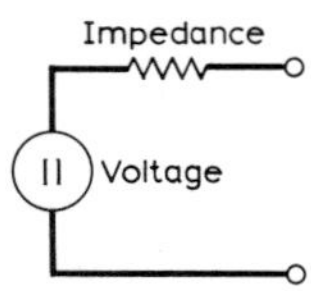

Fig. 9.2 Equivalent circuit of power supply.

(4) the smoothing, usually quoted as the percentage ratio of r.m.s. alternating voltage to d.c. voltage at the output of the power pack. For many purposes this must be held at a very low value, perhaps 0·1% or better.

The performance of the simple power pack of Fig. 9.1 is in many ways unsuitable for any but the most simple pieces of equipment. The off-load voltage is high and the regulation poor. Further disad-

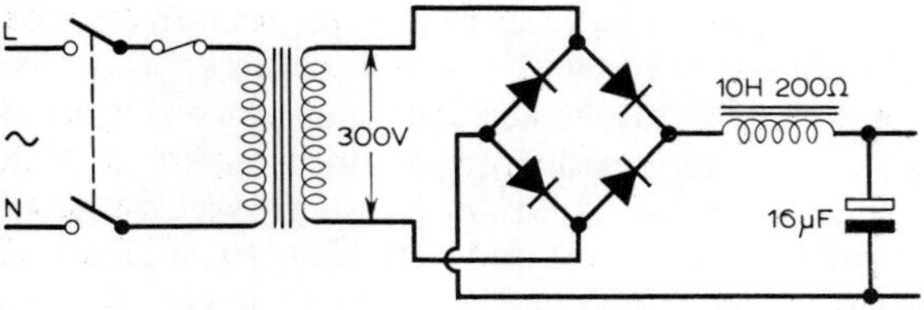

Current out	Voltage out
0	270V
100mA	250V
200mA	230V

Fig. 9.3 Bridge rectifier circuit. The PIV across each diode is only 420V in this circuit; the average current is equal to the average output current.

vantages are that the transformer requires a double secondary winding and that very high current flow into the reservoir capacitor, C, each time a rectifier starts to conduct.

The circuit of Fig. 9.3 offers some improvement, the transformer requiring only one secondary winding and the use of the inductor reducing the high peak current. Because of this, the regulation is better.

Low current high voltage supplies may use the half-wave circuit of Fig. 9.4. The poor regulation of this circuit does not matter much in the type of circuit for which it is used. For higher voltage outputs, one of the voltage multiplier circuits of Fig. 9.5 may be used; again, the regulation is very poor and becomes worse as more stages of multiplication are used.

Failure of Simple Power Packs

Transformer failure may be to open circuit, so that there is no output voltage (checked by measuring alternating voltage of secondary) or to short circuit, sometimes of a small circuit causing overheating

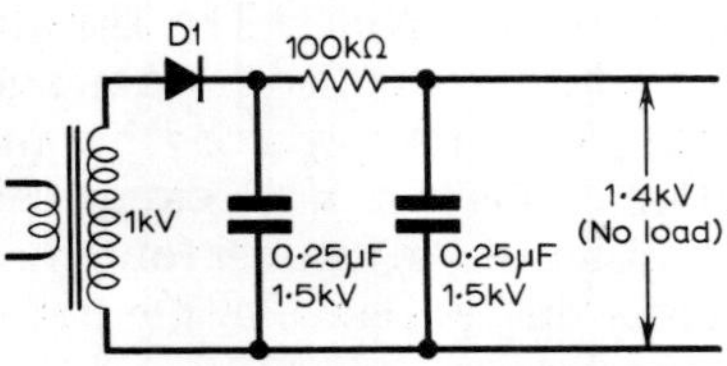

Fig. 9.4 (left) Half-wave rectifier, shown here supplying a high-voltage, low current, output to a c.r.t.

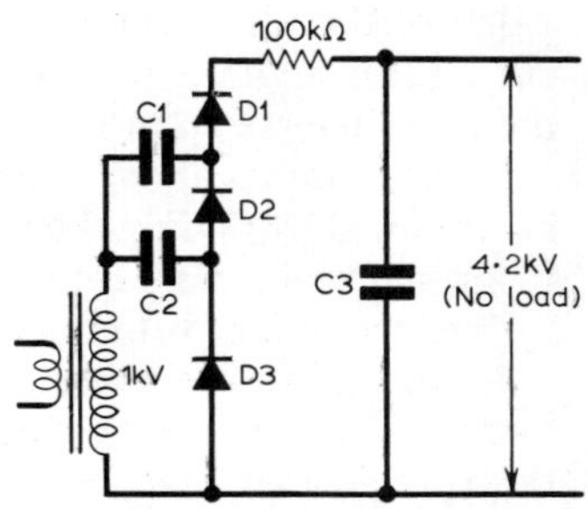

Fig. 9.5 (right) Voltage tripler. This can be extended to any number of stages of multiplication by adding more diodes and capacitors in the network connected to the transformer.

and a drop in output voltage, sometimes of a larger section causing severe overheating and usually blowing power line fuses.

Rectifier failure may be due to open circuit, causing no output in half-wave circuits and severely reducing output along with high ripple in other circuits; or to short circuit causing overloading of capacitors (particularly electrolytics) and fuse blowing. Failure of smoothing components may cause either excessive current through the rectifier when the components short, or excessive ripple when the components go open circuit.

When valve rectifiers are used in a power pack, care should be taken that the rated voltage between heater and cathode is not exceeded. In power supplies which use directly heated valves, the heater supply usually comes from a separate transformer whose primary-to-secondary insulation is sufficient to stand off the fully rectified voltage.

Power Pack Design

The performance of any given circuit at different values of load current can be calculated if the voltage drop across the rectifier at different currents is known. Other factors which must be known are the resistance of transformer windings and the behaviour of any filter circuits. It is convenient to separate the behaviour of rectifier circuits from that of filters, and to work out the performance of each separately. Once the components for the power pack can be specified, allowing for some voltage drop in the filter, then the filter design can be tackled.

Taking first the simplest but least usual case of a half-wave rectifier feeding only a resistor or inductor (which may be part of a filter circuit), the steady voltage out is 0·45V, where V is the r.m.s. value of the alternating voltage input to the rectifier. The average current through the rectifier is equal to the d.c. rectified current, and the peak inverse voltage across the rectifier is 1·4V, where V has the same meaning as above. The regulation of this part of the circuit can be obtained by finding the voltage drop across the rectifier (obtained from manufacturer's tables) and across the resistance of the transformer secondary for any given value of current.

In the more common half-wave rectifier with a capacitive (or battery) load, the peak current is limited only by the resistance of the transformer secondary and the rectifier, and will equal V_p/R_t, where V_p is the peak alternating voltage and R_t the total resistance. This peak current is usually large, but the average current is again equal to the d.c. output current. The peak inverse voltage across the rectifier is 2·8V, double the previous case, because there is an a.c. voltage on one side of the rectifier and a d.c. level, due to the storage in the capacitor, on the other.

Use of the full-wave bridge circuit eases the conditions of rectifiers. With a resistive load, the output voltage is 0·9V, the average current in each rectifier is $\frac{1}{2}I$, where I is the output current, and the peak inverse voltage across each rectifier is 1·4V, with a peak current of 1·57I. Here again, V is the (r.m.s.) a.c. input from the transformer secondary. When calculating the regulation of this circuit, remember that the current is always passing through two rectifiers in series.

The use of an inductor as a load or in a choke-input filter eases the peak current requirements on the rectifier, as the peak current is then no more than the peak a.c. voltage divided by load resistance. Other factors are as for the resistive load. With a capacitive load, the peak current is once again limited only by the total resistance in circuit. The peak inverse voltage across each rectifier is again 1·4V, and the average current is also unchanged at $\frac{1}{2}I$.

The 'full-wave' circuit (more correctly, biphase half-wave) again requires rectifiers of 1·4V peak rating, where V is the (r.m.s.) a.c. input measured between the transformer outputs and not to the centre tap. Average current is $\frac{1}{2}I$ in each rectifier, and only the peak current is affected by a change in the type of load. As before, the peak current into a capacitive load is limited only by the series resistance, and the inductive load has the lowest peak loading on the rectifier, equal to the d.c. output. For a resistive load, the peak current is again 1·57I.

Filter Design

Any filter presents design difficulties, and power supply filters are no exception. In most cases, the aim is to calculate approximate values only and to ensure that a good margin of performance is available. The formulae used here are set out for use on 50Hz mains.

Taking the inductive input filter first, as this has the advantages of better regulation and lower peak rectifier currents, the value of inductance must first be calculated. The value used must be high enough to ensure that the filter behaves as a true inductive input filter. A reasonable approximation to a good value is given by $L1 = R_L/1{,}000$ henries, where R_L is the load resistance presented to the other side of the filter by the load, and is equal to $V_{out}/I_{out} \cdot R_L$ must be in ohms, so V_{out} should be in volts and I_{out} in amps. This value of inductance should be regarded as the safe minimum for the filter first stage.

The product of $L1$ and $C1$ for the first portion of the filter can now be worked out in terms of the amount of a.c. ripple which can be tolerated on the output. If this ripple voltage is V_r volts, then $V_r/V_{out} = 1{\cdot}2/L1C1$ for a full wave rectifier circuit working at 50Hz. From this it can be seen that the amount of ripple voltage obtained depends on the value of $L1C1$ used. In the calculation $L1$ should be in henries and $C1$ in microfarads.

For further ripple reduction, a second filter section can be used, and will give a further reduction in ripple depending on the value of LC constant used. If the ripple voltage from the second section is to be V_{rr}, then $V_{rr}/V_r = 2{\cdot}5/L2C2$, where $L2$ and $C2$ are the inductance and capacitance values used in the second section, again in henries and microfarads.

When a capacitor input filter is used, the CR product for the ripple obtained is given by: $V_r/V_{out} = 4{,}500/C1R1$, where $C1$ is in μF and $R1$ in ohms. The value of $C1$ is selected so that the peak current of the rectifier (manufacturer's maximum working value) is not exceeded when it is loaded by $C1$. This condition is satisfied if the value of $C1$ in μF is less than the amount given by $(I/V) \times 2{,}350$, where I is the peak allowable current in amps and V is the r.m.s. voltage applied to the rectifier.

If more smoothing is required a second section of RC filter can be added, or a second section of LC filter with the same order of ripple reduction. Care must be taken that the L and C values used in filters do not resonate at 50Hz, so causing dangerously high currents to flow. Values of LC equal to about 10 (L in H, C in μF) are to be avoided.

Stabilisers

The voltage of a simple power pack drops as current is taken from it, as is the case with a battery, and for many purposes this is undesirable; it can cause unwanted feedback in amplifiers, misfiring in pulse circuits and inaccuracy in measuring instruments, all of which can be avoided by the use of a stabilised supply. An ideal stabilised supply would give a steady voltage unaffected by the current drawn within the limits of its ratings and also unaffected by variations in the mains voltage of up to 20% or so.

Perfection is not attainable, but very satisfactory stabilisation of voltage, and current if required, can be obtained by circuits which, although they appear complex, are made up of portions already familiar from earlier chapters. The most used technique is to obtain a low-voltage stable d.c. source and amplify it, using a d.c. amplifier with feedback arranged so that the output impedance is low. This is not a conventional way of looking at a stabiliser circuit, but it has the merits of simplicity and enabling the circuits to be seen as familiar patterns.

Since a stable voltage is being amplified, the output should be stable despite mains fluctuations unless these fluctuations cause large changes in the amplification which cannot be completely hidden by the action of the feedback. Also because of the feedback, the output impedance is low, causing the voltage drop when the current is varied to be small. Output impedances of the order of 0·01Ω are obtainable.

For such a stabilising amplifier to be used, some source of stable voltage must be available. Ideally, this voltage should be about the same as the voltage to be supplied, so that little amplification is needed, but in practical cases smaller voltages have to be used. The source of stable voltage, the reference source, could be a primary cell (which would have a very long life, as very little current would be taken from it) but is more usually a gas stabiliser or a zener diode.

Gas Stabilisers

A gas stabiliser is a valve-like device which uses a metal rod cathode with a metal cylinder anode fitting over the cathode but not in contact with it. The cathode is not heated, and the valve is filled with a gas such as Argon, Neon, Krypton or Xenon at low pressure.

Such an arrangement conducts electricity only at voltages which cause the gas to ionise, each atom separating into electron and ion, and the behaviour of the ionised gas is not at all like that of a resistance. An increase in current tends to cause an increase in the number of ions (caused by collisions between ions and atoms) and so the voltage across the valve drops slightly instead of rising as would be the case with a resistor. Any device whose voltage/current graph is not a straight line is called non-linear, and if the voltage decreases as the current increases, then the term 'negative resistance' can be used, as this is the opposite of normal resistor behaviour. Most gas stabilisers therefore behave as non-linear negative resistors.

The graph of voltage and current for a typical gas stabiliser is shown in Fig. 9.6, and we can see that no current passes until the voltage reaches a peak called the striking voltage. When current passes, the voltage across the stabiliser drops to another value, the running voltage, which is the value of stabilised voltage obtained.

Different combinations of gases and cathode metals can be used to give a range of valves with stable voltages of 45V, 60V, 85V, 90V,

105V, 150V and 180V standardised. Gas stabilisers can be connected in series to give higher voltages, but parallel connection must not be used; it has no advantages, and slight differences between running voltages cause one stabiliser to carry an unfair amount of the current.

A resistor must always be connected in series with a gas stabiliser to limit the amount of current which can pass, otherwise a 200V supply connected to a 105V stabiliser would cause all the system fuses to blow, as the voltage across the stabiliser could not rise sufficiently above 105V to equal the applied voltage, even when the gas became fully ionised and voltage started to increase with increasing current.

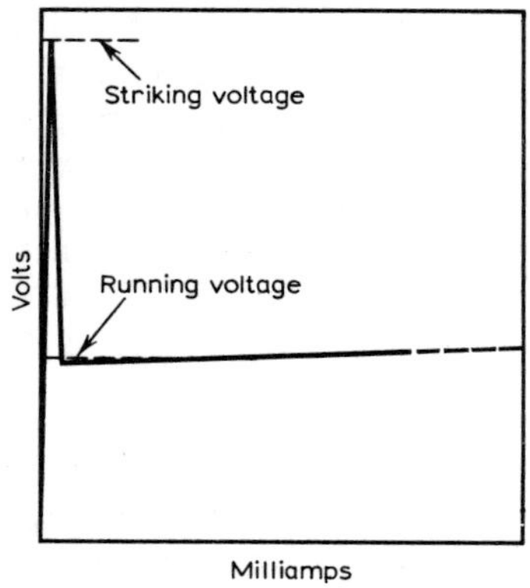

Fig. 9.6 Gas stabiliser characteristic.

Gas stabilisers are specified by their running voltage, striking voltage, variation of voltage output with current drawn (if this is particularly small, then the valve is known as a 'reference tube') and maximum operating current. In some cases, the gas is made slightly radioactive to assist in parting electrons from atoms, so reducing the difference between striking voltage and running voltage; this has also the effect of making the value of the striking voltage less dependent on the amount of light reaching the tube, since light has the same effect on the gas as radioactivity, and a non-radioactive gas causes the tube to have much higher striking voltage in the dark than it has in light.

Zener Diodes

The zener diode is the semiconductor counterpart of the gas stabiliser. When a reverse voltage is applied to a diode, little current flows until the voltage reaches a value, the avalanche voltage, at which the

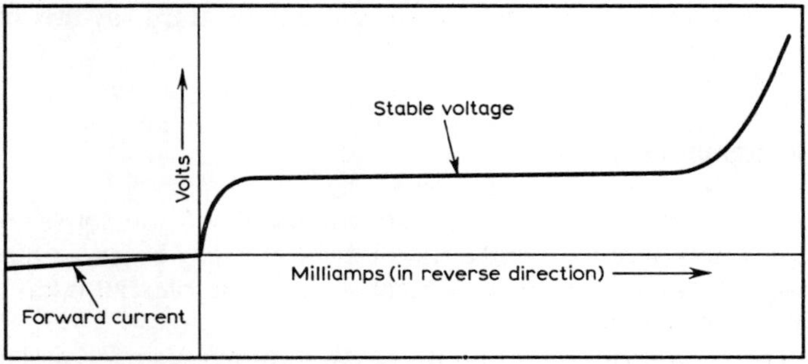

Fig. 9.7 Zener diode characteristic. Note that the current scales have been drawn in a direction opposite to that of the usual diode characteristic.

current rises very sharply for any small increase in voltage. Diodes can be made in which the increase in current is very sharp and occurs at a precise voltage. Such diodes should be called avalanche diodes but are universally called after Clarence Zener, who discovered another similar effect appearing at higher voltages.

The voltage-current graph for a zener diode is shown in Fig. 9.7; note that the graph for forward voltages is quite normal, so that a zener diode can be used as a normal diode as well as for stabilising (unlike a gas stabiliser). Note also that there is nothing corresponding to the striking voltage/running voltage difference.

Zener diodes can be made with a very wide range of stabilised voltages, a much greater range than is possible with gas stabilisers. They also behave like resistors in the sense that increase of voltage always causes an increase of voltage across the diode; there is no region of negative resistance. Zener diodes are specified by their stable voltage and percentage tolerance, the 'dynamic resistance', the temperature coefficient of voltage and the dissipation. The dynamic resistance is equal to

$$\frac{\text{change of voltage}}{\text{change of current}}$$

for a small change of current while the diode is working, and is in ohms; it is the impedance of the diode.

The temperature coefficient of voltage of the diode measures the percentage change in stable voltage caused by an increase of 1°C in temperature, and can be positive (when the voltage increases as temperature increases), or negative (when the voltage decreases as the current decreases), depending on the design of the diode. The dissipation, in watts, enables the maximum working current to be calculated. For example, if a 1W diode stabilising at 5V is used, then

$1 = 5 \times I$, where I is the maximum current in amps, in this case 1/5 A or 200mA.

Simple Stabiliser Circuits

The simplest stabiliser circuits are shown in Fig. 9.8 and consist of a voltage supply, a resistor and a stabiliser, which may be gas or zener. For some purposes this simple circuit is quite suitable, but it has the following drawbacks:

(1) The voltage out is the voltage of the stabiliser element, and can be changed only be changing the element.

(2) If no current is drawn by the load during some part of its operation, then the load current will flow to the stabiliser which may cause it to exceed its dissipation. The maximum current of the stabiliser element, therefore, is the maximum which can be allowed to be used by the load, since the load takes current at the expense of the stabiliser.

(3) The circuit is wasteful in the sense that the same current has to flow through the dropping resistor whether or not the load is taking current. This resistor must continually dissipate heat.

Despite these disadvantages, the simple method is used extensively, especially when a stable voltage is required for only one stage of a circuit. Before we examine more complex stabilising methods, we can compare the performance of gas stabilisers and zener diodes in the simple circuit.

(1) At stable voltages up to 45V, the zener diode must be used, since lower voltage gas stabilisers cannot be made. At voltages of 75V and over, zener diodes are very much more expensive than gas stabilisers.

(2) Where gas stabilisers are used, the unstabilised voltage must be considerably more than the striking voltage of the valve. Striking may be aided by using a stabiliser with an 'ignition electrode', an extra anode which can be used to cause triggering at a voltage closer

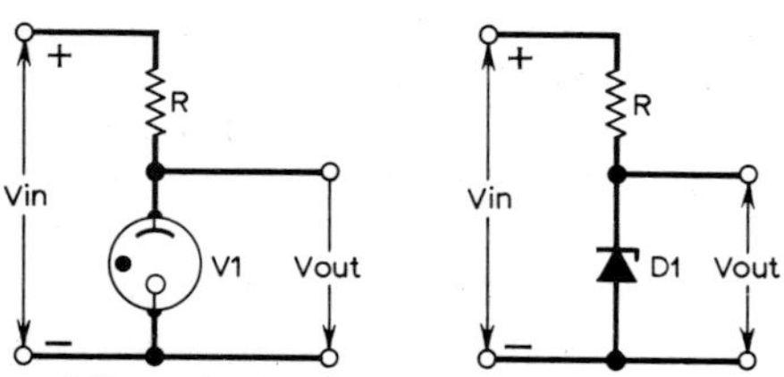

Fig. 9.8 Simple stabiliser circuits. In each case the voltage across the resistor is (Vin−Vout), and the wattage dissipated in the resistor is (Vin−Vout). Iin.

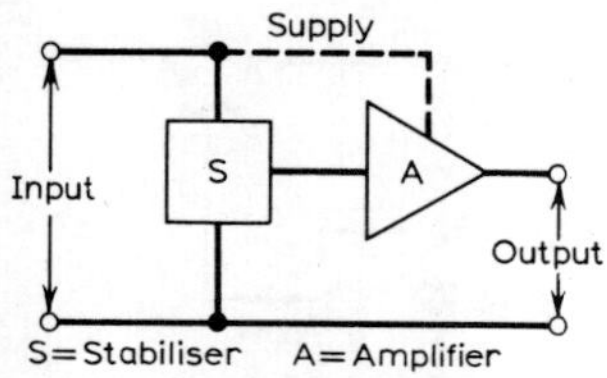

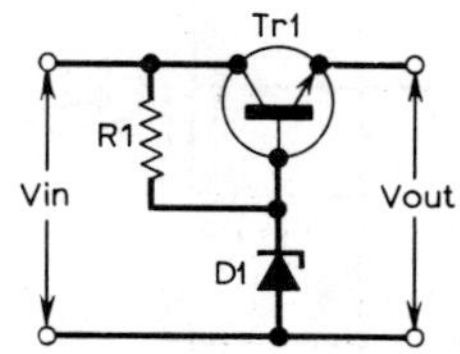

Fig. 9.9 (left) Block diagram of stabilising or regulating circuit.

Fig. 9.10 (right) Simple transistor stabiliser circuit.

to the running voltage, but the difference between striking and running voltages always exists.

(3) If the unstabilised voltage goes momentarily below the stabilised voltage, a gas stabiliser extinguishes and ceases to conduct, and does not pass current again until the unstabilised voltage has risen above the striking voltage. When this happens with a zener diode, control is regained smoothly whenever the unstabilised voltage rises above the stable level again.

(4) The gas stabiliser behaves as a negative resistance while it is running. This behaviour is also shown by an amplifier with positive feedback, and any kind of negative resistance will oscillate if connected to a time constant. For this reason, care has to be taken if a time constant is attached to a gas stabiliser, and the addition of a capacitor to the existing series resistor is enough to constitute such a time constant. Oscillation is not always inevitable, since the system has a low 'Q' value, but no such precautions are needed with zener diodes.

(5) The zener diode is much smaller physically than the gas stabiliser.

More Advanced Stabiliser Circuits

We mentioned earlier that the most used stabiliser circuits consisted of a stabilising element connected to a d.c. amplifier with negative feedback used to keep output impedance low and amplifier conditions stable. In outline, such a circuit would appear as in Fig. 9.9, and this outline can be used as a guide to any voltage stabilising circuit.

A very simple circuit, in which the d.c. amplifier is not used, is shown in Fig. 9.10. In this circuit, the voltage at the base of the transistor is set by the zener diode. Any drop in voltage at the load causes a drop in voltage at the emitter, thus more current flows to

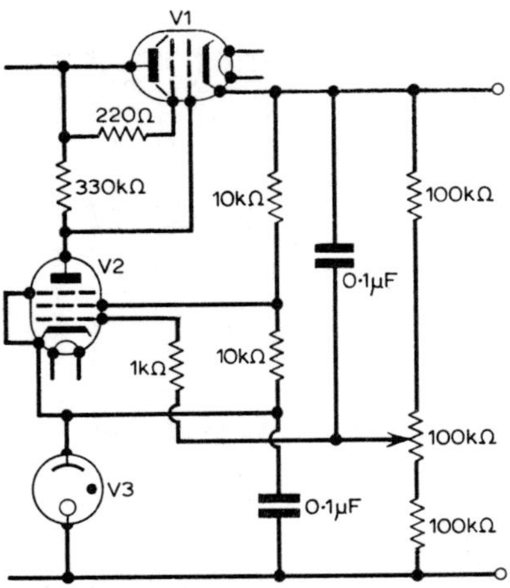

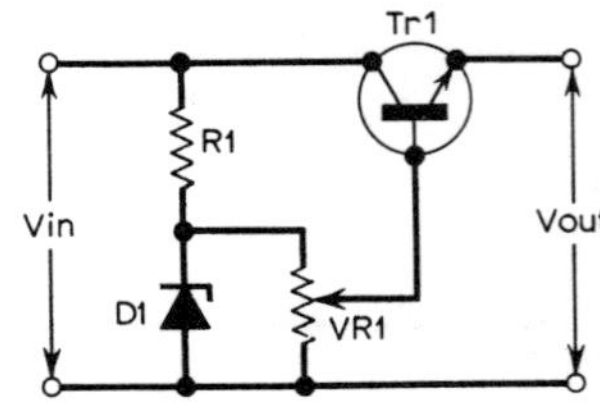

Fig. 9.11 (above) Simple variable-voltage supply.

Fig. 9.12 (left) Valve stabilised circuit. V1 is a power pentode, V2 a general purpose voltage amplifier pentode.

the base and greatly increased current flows to the collector. This increased current flowing into the load restores the load voltage to nearly normal, so achieving stabilisation. The transistor is acting as a current amplifier, and in this way we remove the second disadvantage of the simple stabiliser circuit of Fig. 9.8.

We can remove the first disadvantage (fixed voltage operation) by the modification shown in Fig. 9.11, where the stable voltage of the zener diode is applied across a voltage divider to the base of the transistor, so that a variable stable voltage can be applied if the voltage divider is of low impedance. The third disadvantage of the simple circuit is also removed because when no current flows through the load, none flows through the transistor, and the only current left flowing is that through the zener diode, which can be small, as it does not have to feed the load.

The negative feedback in this circuit is applied by using the transistor as an emitter follower, but no voltage amplification has been used; the output voltage is always rather less than the voltage of the zener diode. The circuit shows, however, most of the features of more complex stabilisers: the voltage at the output is almost independent of variations of the unstabilised voltage at the input because of the transistor being controlled by the voltage at its base.

The output voltage is also less affected by variations in the load current: a typical transistor might have a voltage between base and emitter of 0·1V at 100mA and 0·8V at 1A collector currents, so that the output voltage would vary by only 0·7V for a current change of 900mA. This gives an output resistance of 0·7/0·9, which is 0·78Ω compared to a probable value of 20Ω for the unstabilised circuit.

This circuit can also be used with a valve but is less effective as a stabiliser.

Stabiliser Circuits

A typical valve stabilised circuit is shown in Fig. 9.12; it uses two valves and a gas stabiliser element. The gas valve sets the voltage at the cathode of *V*1 to 150V, and carries the cathode current of *V*1 as well as the current through the two 12kΩ resistors which keep the gas valve alight even if *V*1 cuts off. The grid of *V*1 is fed by the potential divider *R*3, *R*4, *R*5 which delivers a fraction (adjusted by *R*4) of the regulated voltage at A to the grid. If this voltage should rise, the grid voltage rises (cathode is held constant), *V*1 draws more current and the voltage at B drops. This biases back *V*2 which then passes less current until the voltage at A is restored to normal.

*V*2 behaves as a cathode follower whose cathode voltage closely follows its grid voltage. If the voltage at A falls, the reverse process happens. This fairly simple regulator can give a good performance, but for still better regulation or wider range of output voltage more complex circuits have to be used.

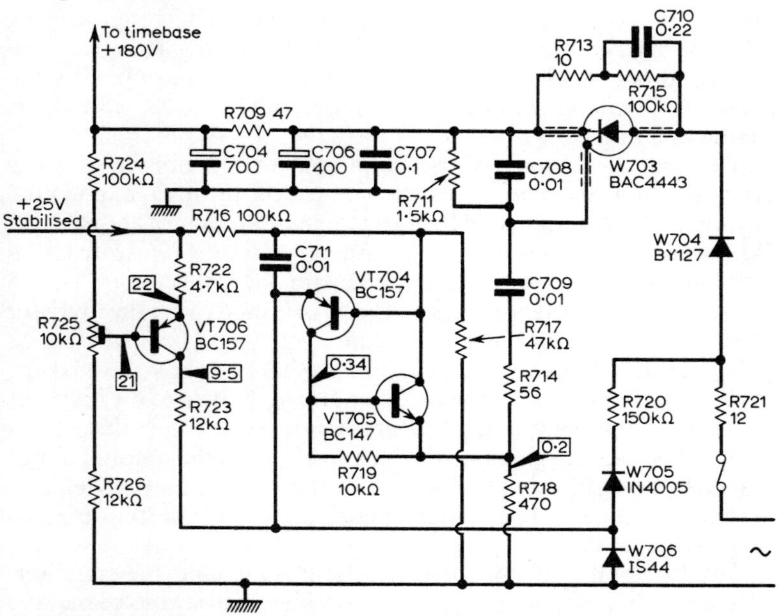

Fig. 9.13 Thyristor stabiliser as used in the BRC8000 colour TV chassis. Steady bias voltages are shown within rectangles.

Thyristor Stabilisers

Another method of stabilising voltage uses thyristors to overcome the problem of large steady dissipation in power transistors or valves, with the added advantage that thyristors can work at voltage levels for which previously only valves were suited. The principle behind thyristor stabilisers is fairly straightforward; the output voltage is sampled and used to control the phase angle at which a thyristor fires. The earlier in a half wave that the firing occurs, the greater the voltage output of the rectifier of which the thyristor is part, and so if the phase angle is controlled by the output voltage the negative feedback loop needed is complete.

Fig. 9.13 shows the circuit used for the e.h.t. stabilisation of the BRC 8000 series colour TV receivers. *C*712 is a capacitor which is charged from a stabilised +25V line through *R*722 and *R*723. The current available to charge the capacitor is controlled by the p-n-p transistor *VT*706, and is set by the base input which is derived from the +170V supply through *R*724, *R*725 and *R*726.

The 170V supply is stabilised by the thyristor, so it is this supply which is being compared with the stabilised 25V supply in the transistor *VT*706. If the 170V supply is low in voltage, making the base of *VT*706 more negative, the charging current is greater, causing *C*712 to charge more quickly. If the 170V supply is high, *C*712 charges more slowly. *C*712 is discharged every mains cycle by the mains voltage applied through *R*720 and diode *W*705, with *W*706 acting as a limiter.

The charging supply is also connected to the emitter of p-n-p transistor *VT*704 whose base is connected through a potential divider to the 25V stabilised line, giving a base voltage of about +8V. As the emitter voltage increases to above +8V, *VT*704 switches on and also switches on *VT*705. Current in the collector of *VT*705 causes a drop of voltage across *R*716 which is used to fire the thyristor *W*703 through *R*714, *C*709 in the gate circuit.

The feedback between *VT*704 and *VT*705 is positive so that the switchover is very rapid. The current pulse to the gate of the thyristor fires it, and the point in the forward half wave at which this occurs depends on the charging rate of *C*712 and thus on the output voltage obtained by smoothing the voltage at the thyratron cathode. This stabilised voltage is then used to operate the TV line output stage, so generating the e.h.t.

Thyristor stabilising can also be used as a valuable 'pre-stabiliser' in a stabilised power supply, coping with mains variations so that the main stabiliser is concerned with presenting a very low impedance to the circuit being fed.

Fig. 9.15 shows a stabiliser circuit of a more advanced type which uses the output of a thyristor stabiliser to power a series stabiliser circuit. The series stabilising transistors, *VT*6 to *VT*13, are all connected in parallel. Transistors *VT*3, 4 and 5 form a current amplifier so that small current changes in the base current of *VT*3 provide enough current in *VT*5 to drive the series stabilisers *VT*6 to *VT*13, all *OC*28's. *VT*16 and *VT*17 form a long-tailed pair; *VT*16 has the zener diode *ZD*1 in its base circuit and *VT*17 has a fraction of the output voltage selected by the range switch Sw2(e) applied to its base.

*VT*18 amplifies the output of *VT*17 before applying it to the base of *VT*3. *VT*14 and *VT*15 form an overload circuit which switches off current when a limit, set by Sw3, is reached. This power supply has a voltage range of 0–50V, a current range of 0–10A and an output impedance of 0·001Ω. Variation of output voltage caused by mains voltage variation is typically 1/3000 of the mains voltage variation.

Integrated Stabilisers

The comparator circuits of a stabiliser are identical in pattern to the balanced stages used in op-amps, so it is not surprising that voltage regulators are available in integrated form. Generally these incorporate their own zener diodes, with provision for externally varying the stabilised voltage, but several designs provide for an externally connected zener diode so that the output voltage levels can be raised.

Because of the limitations of power dissipation in integrated circuits the current capabilities of the regulators are small, but the basic regulator can be used to control a power transistor. Fig. 9.14 shows the RS Components (Radiospares) regulator used along with a 2N3055 transistor to regulate a 2–30V, 0–1A supply. This regulator incorporates short-circuit protection, and the stabilised current

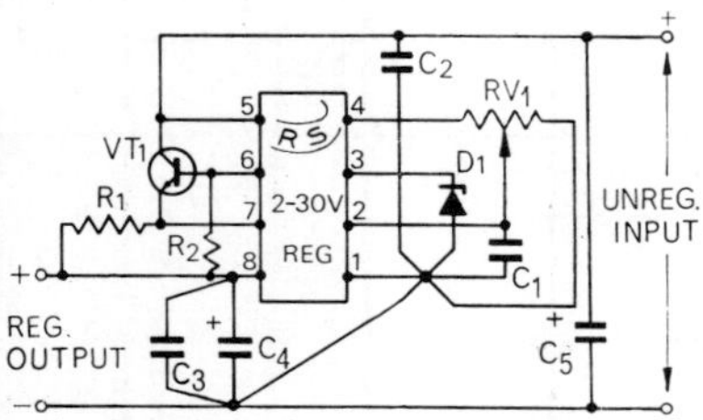

COMPONENTS

R_1	·47Ω 1W W/W Resistor
R_2	470Ω ½W Resistor
C_1, C_2, C_3	·1µF 250V Polyester
C_4	100µF 50V Tube
C_5	3300µF 63V H-R Electro
D_1	M-ZE 8·2V
2—30V Regulator	
RV_1	5kΩ Midget Lin:
VT_1	2N3055

Heat Sink for VT1
(Mount with smear of silicone grease)
Keep wires as short as possible and wire as shown to minimise hum.
Heat Sink for 2—30V Regulator:—
Size 50 mm × 50 mm × 18 s.w.g.
Bright Aluminium.

Fig. 9.14 IC regulator used with external transistor. (*courtesy of RS Components*)

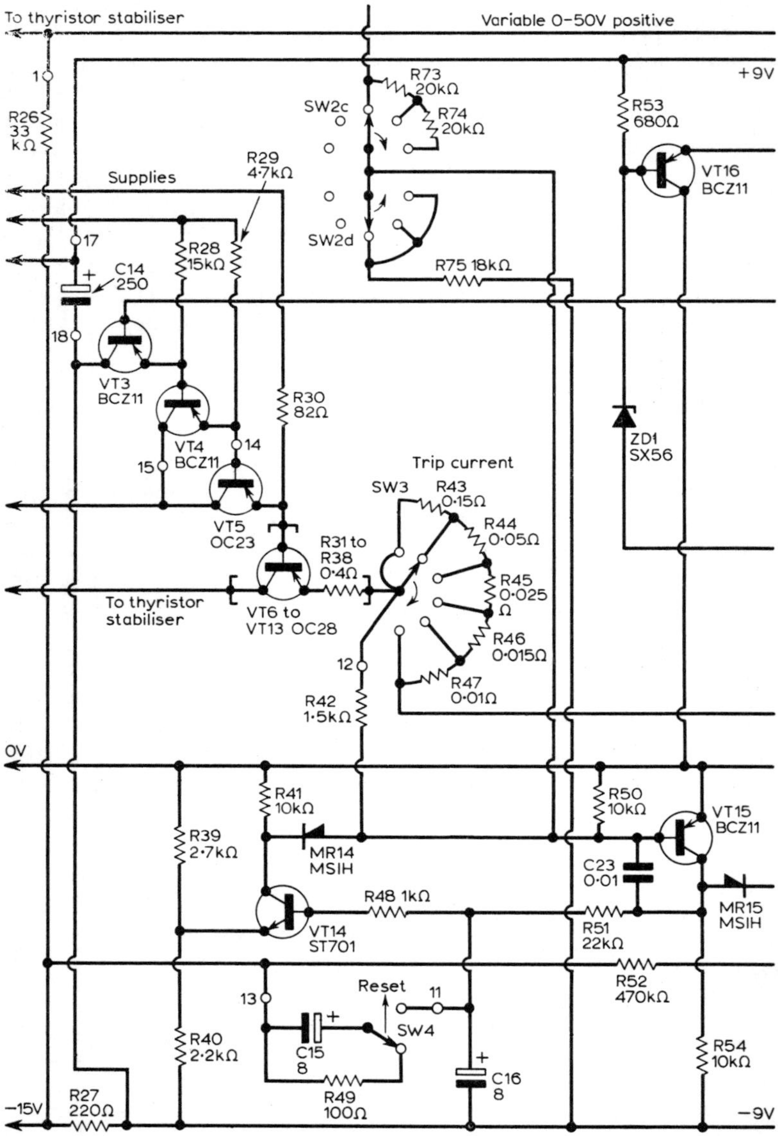
To thyristor stabiliser
Variable 0–50V positive
+9V
1
R26 33 kΩ
R73 20kΩ
SW2c
R74 20kΩ
R53 680Ω
VT16 BCZ11
R29 4·7kΩ
Supplies
17
R28 15kΩ
SW2d
C14 250
R75 18kΩ
18
VT3 BCZ11
VT4 BCZ11
R30 82Ω
ZD1 SX56
14
15
Trip current
SW3
R43 0·15Ω
VT5 OC23
R44 0·05Ω
R31 to R38 0·4Ω
R45 0·025 Ω
To thyristor stabiliser
VT6 to VT13 OC28
R46 0·015Ω
12
R47 0·01Ω
R42 1·5kΩ
0V
R41 10kΩ
R50 10kΩ
VT15 BCZ11
R39 2·7kΩ
MR14 MSIH
C23 0·01
R48 1kΩ
MR15 MSIH
VT14 ST701
R51 22kΩ
R52 470kΩ
Reset
13
11
SW4
R40 2·2kΩ
C15 8
R54 10kΩ
C16 8
R49 100Ω
R27 220Ω
−15V
−9V

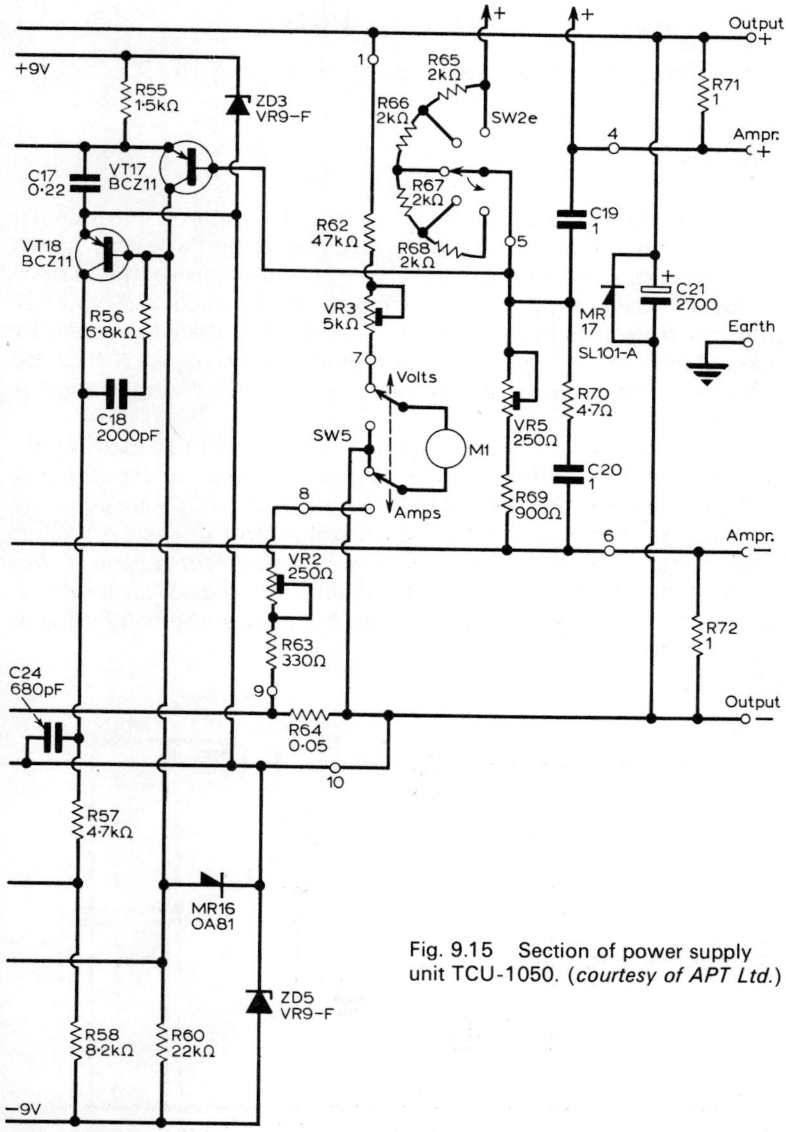

Fig. 9.15 Section of power supply unit TCU-1050. (*courtesy of APT Ltd.*)

obtainable depends on the setting of the s/c protection network. The stabilisation ratio

$$\left(\frac{\text{change of } V_{in}}{\text{change of } V_{out}}\right)$$

is typically 225 and the output impedance typically 0·15Ω.

Inverters

In many cases, voltages which are required cannot be obtained conveniently either directly from batteries or by the use of mains power packs. For example, portable transmitters using valves need a high voltage supply which cannot be obtained conveniently directly from a battery, and e.h.t. supplies of 10kV upwards for cathode-ray tubes require large and bulky power packs if they are obtained from a mains power pack, particularly if stabilisation is required.

In such cases, the circuit technique of inversion is used, as indicated in Fig. 9.16. D.C. is used to power an oscillator working at a high frequency, and the oscillator voltage output is stepped up by a transformer, which can be a small component if it is working at high frequencies. After this process of inversion (conversion of d.c. to a.c.) the output can be rectified and, if necessary, stabilised. Stabilisation can be easily carried out by using a negative feedback loop operating on the oscillator.

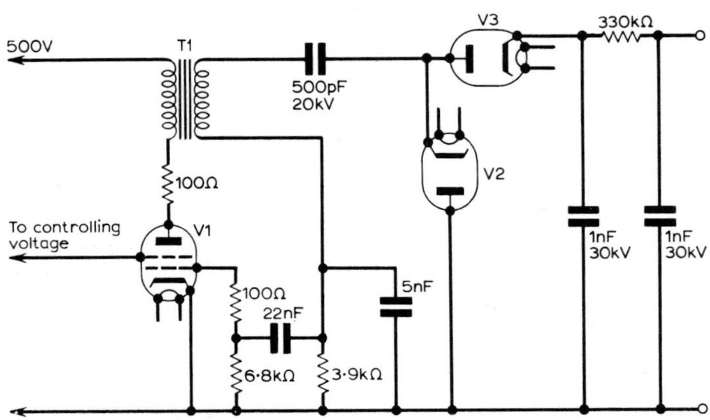

Fig. 9.16 High voltage supply using inversion. This is a simplified portion of a stabilised 30kV generator. (*courtesy of Brandenburg Ltd.*)

Inverters are also found changing d.c. of various voltages into mains voltage and frequency a.c., sometimes with elaborate circuitry to protect against overload, sometimes with arrangements to switch into a load whenever a mains failure occurs. Small inverters of this type use power transistors as the oscillators, often in a cross-coupled multivib type of circuit but with the primary of a large transformer as the collector load of the push-pull circuit.

Larger inverters, now becoming very common as each winter brings fresh power supply troubles, use thyristors to commutate (switch-over) the currents in the primary windings of the transformer. Output powers vary from about 10W for the inverters used for caravan fluorescents to more than 10kW for the emergency inverters for computer installations.

CHAPTER TEN

THE OSCILLOSCOPE

FOR ANY SORT OF SERIOUS WORK in electronics, whether servicing, modifying, testing or designing, an oscilloscope is now quite indispensible, but unfortunately it is not always used as constructively as it might be. This chapter is a brief guide to yet another topic which deserves a book of its own, the operating principles, use and servicing of the oscilloscope.

Basically, the oscilloscope presents a graph in which voltage is measured vertically and time is measured horizontally. Anything which can be converted into voltage, such as current, acceleration, velocity, light intensity, temperature level, pressure, etc., can be displayed on the vertical axis, and anything capable of being converted into a voltage can be displayed on the horizontal axis as well as the built-in time scale which is available.

Oscilloscope Fundamentals

Fig. 10.1 shows a block diagram of an oscilloscope and a list of some of the controls commonly found. The c.r.t. is the central feature of the oscilloscope, and its function is to project a beam of electrons on to a screen which detects the electrons by giving out light where it is struck. The intensity of the beam (causing display brightness) may be controlled by the *G*1 voltage of the electron gun. The position of the spot formed when the beam is focused may be varied by the deflection plates, the X-plates controlling horizontal movements and the Y-plates controlling vertical movements.

Each of these sets of plates is driven by a voltage amplifier whose peak output, applied so that the plates are driven in push-pull, is sufficient to deflect the spot well beyond the edge of the screen. In some cases the peak output of the amplifier may be sufficient to drive the spot to several screen diameters, and the result is that any trace can be expanded so that a trace can be 'blown up' to examine

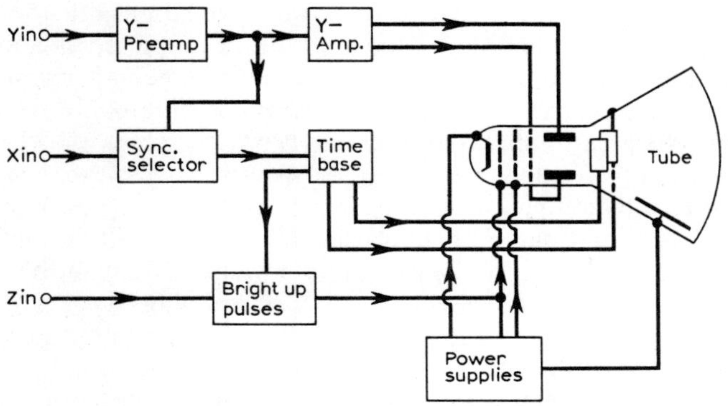

Fig. 10.1 Block diagram of oscilloscope.

detail in a small portion of a waveform. Trace expansion is frequently used on the amplifier which supplies the X-plates so that time amplification of this sort can be performed without affecting the synchronisation of the trace, as would be the case if the speed of the trace were altered.

The X-expansion control is a gain control on the X-amplifier, and some control will be found on the Y-amplifier also, this time labelled 'Y-gain' in volts/cm, and often a switched control rather than a continuous (potentiometer) control. The gain control will generally be calibrated, or can be calibrated by comparison with a standard signal, in terms of the number of volts required at the input of the oscilloscope to deflect the trace on the screen vertically by one cm.; the accuracy of this calibration is important if the oscilloscope is to be used as a measuring instrument as distinct from a means of viewing waveforms.

The Amplifiers

Both X- and Y-amplifiers are usually direct-coupled to the plates of the oscilloscope so that the plates are driven in paraphase, and the shift controls form part of the amplifier. The shift controls enable the undeflected position of the trace to be shifted to any part of the tube face so that the whole of a trace can be displayed, or the voltage levels of traces compared, or times measured, in each case by examining the waveforms through a ruled screen (graticule) and measuring the size of the waveform.

As well as the Y-amplifier, there may be a Y-preamplifier. Very often, the main amplifier is direct-coupled and has its gain and bandwidth fixed, the variations of gain being brought about by using a switched attenuator at the input of the amplifier. In addition to this, the oscilloscope may use a preamplifier for smaller signals so that higher sensitivity at lower bandwidth is obtainable. In some designs, the preamplifier is automatically switched into use when the range switch is rotated to the higher sensivity ranges; the preamp is very often CR-coupled so that small d.c. levels cannot be compared.

The labelling of the V/cm control is often arranged to remind the user of the limitations of the preamplifier in use. Many of the larger and more complex oscilloscopes use a system of plug-in preamplifiers so that a choice of high-gain/low-bandwidth, low-gain/high-bandwidth, balanced or d.c., preamplifiers may be used in conjunction with the very wideband main Y-amplifier. Since there is a preamp for each purpose, fewer design compromises have to be accepted, and a much higher standard of performance can be achieved.

Timebase Output

The X-amplifier may be driven externally, and on many 'scopes there may be a position of the 'sweep time' or 'speed' switch which couples the input of the X-amplifier to an external socket, often at the rear of the case. For most work, however, the X-amplifier will be driven by the timebase. The output of the timebase section is a sawtooth (linear ramp) waveform which has the effect of deflecting the spot from one side of the screen to the other at an even speed during the time of the sweep, and back again at a much higher speed; the backstroke is known as the flyback.

An additional output from the timebase section is the blanking pulse which biases back the grid of the c.r.t. during flyback; this may be supplemented by a 'brightup' pulse which switches the beam on during the forward stroke, or the brightup pulse alone may be used. Both pulses are desirable, because the brightup is most effective at high sweep speeds and the blanking at low sweep speeds. The reason for this is that the pulse has to be coupled to the c.r.t. grid by a time constant, unless an elaborate form of d.c. level change can be attained (the grid will be about 2000V negative to earth), and long pulses, such as the brightup on a long sweep, tend to be integrated.

At the same time, stray capacitance on the coupling limits the very short pulses used to blank a rapid flyback. The aim of both pulses is to ensure that the trace is visible only during the forward stroke; in the case of the brightup pulse, the amplitude of the pulse may be adjusted to compensate for the dimming of faster traces (at fixed bias) due to the smaller time during which electrons hit the screen.

Sweep Speed

The slope of the sawtooth waveform decides how fast the spot will move across the screen, and this slope is usually variable by two controls. One control varies the capacitance in the timebase charging circuit and in the squarewave generator, and is a switch whose labelling is in μS, mS, or seconds of sweep for the maximum time of sweep, or time per cm., available at that setting. The other control is a potentiometer which varies the speed over a smaller range so as to select the precise speed needed to synchronise a waveform.

In some cases, the potentiometer is not calibrated except at the extreme ends of its track. In other 'scopes the potentiometer is fully calibrated as a fraction of the main scale; for example, if the main scale positions are 1μS, 10μS, 100μS etc., the pot. may be calibrated 1–10, so that a main scale setting of 10μS and a pot. position of 5 means a sweep time of 50μS. In expensive instruments, the fine adjustment made possible by the use of a potentiometer may be made finer still by the use of a multi-turn helipot.

Some advanced instruments use, in addition, a second timebase which can be triggered from any point on the first. When this is used, the part of the trace covered by the second (faster) timebase is indicated by an additional brightup of that part of the main trace, and altering the position at which the second timebase triggers will move the position of the brightup.

The second timebase can then be switched into the X-amplifier instead of the first, so showing a greatly time-magnified view of part of the trace. In this type of use, the main timebase is acting as a time delay trigger for the second timebase, and great accuracy of timing and triggering is required to avoid jittering, the random to-and-fro movement of the trace in the magnified form.

Timebase Controls

Also concerned with the timebase section are the controls variously labelled 'sync', 'trigger-level', 'sync. amp', which control the locking of the timebase to the input waveform. To show a steady trace on the screen, it is necessary that the timebase should start at the same point in the display of a waveform each time, and the various controls concerned with synchronisation adjust the conditions needed to achieve that end.

On many modern 'scopes, one of the controls will labelled 'auto', and in this switch position, any signal of the minimum amplitude needed will lock the timebase so that a steady trace is displayed. 'Trigger Level' usually controls the locking amplitude, so that the timebase can be locked to different parts of the waveform, depending on their amplitude. In this way, slight alterations of the phase of the trace relative to the start of the timebase can be made.

'Sync. selector' is a switched control which may provide a bewildering range of choice. This control selects:

(1) The mode of operation of the timebase, whether triggered or free-running. A triggered timebase starts only when a sync pulse is received, and is therefore completely locked to that pulse. The pulse may be taken from an external source, as for example the clock pulse of a computing or counting system, or it may be derived inside the 'scope from the waveform being shown. A free-running timebase runs whether any waveform is being displayed or not; application of sync can alter the running speed slightly so that the waveform is locked, but the lock is not complete. When the fine speed control is adjusted on a free-running timebase, the waveform display slips in and out of lock as the speed is varied; a triggered timebase under such circumstances simply shows the waveform being expanded and contracted horizontally without going out of lock.

(2) The timebase can be made to lock on positive portions of waveforms only or on negative only.

(3) The timebase can take the sync/trigger pulses from the waveform being observed by tapping the waveform from the Y-amplifier and squaring, or it can run from an external source of pulses.

(4) The timebase can be synchronised from the sync pulses of a TV signal, by separating the pulses from the video signal and using them to synchronise the 'scope timebase. Either fields or lines can be shown, but it is not usually possible to show one selected line unless an additional attachment is used, since this involves selecting one line out of each complete frame and synchronising the timebase once per frame on that line. In oscilloscopes designed specially for the servicing of TV transmitting apparatus such a facility may be built in.

Other Oscilloscope Controls

As well as the main controls already described, the oscilloscope will have the following:

'Brightness' or 'Brilliance' control to set the brightness of the trace. If the oscilloscope does not use a bright-up pulse then the brilliance will have to be set low when the spot is not deflected (such as when it is triggered and the trigger pulses are interrupted) and high when a fast trace is being viewed. It is undesirable to have slow traces of high brilliance, since there is a risk of screen damage.

'Focus' enables the trace to be made as uniformly thin and well defined as the e.h.t., the design of the c.r.t. and deflection amplifiers permits. In the simpler 'scopes, really fine traces are not obtainable mainly because of the low voltages used on the tube. On more advanced instruments, hairline traces can be obtained, especially if the focus is assisted by an 'Astigmatism' control. This latter control determines the shape of the scanning spot, particularly at the edges of the screen, and should be adjusted to give a round spot whose size can then be adjusted by the focus control.

A good test of focus and astigmatism controls is to display a trace of closely spaced squarewaves. If focus and astigmatism are correctly adjusted, both tops and sides should appear to have equally thin lines and any adjustment should only thicken the trace. Be sure that your own eyesight is corrected for astigmatism first! The following controls and facilities may also be found:

(1) Graticule brightness—a control which varies the illumination of the graph line pattern used for time and voltage measurement.

(2) Time calibration—an internal high-stability oscillator provides waveforms of known frequencies for checking time measurement.

(3) Voltage calibration—a squarewave of (usually) 1V p-p can be used for checking the accuracy of the amplifier calibrations.

(4) Z-modulation enables a connection to be made to the grid of the c.r.t. by a CR coupling. An external waveform may then be used to bright-up or blank-off portions of a trace.

(5) A.C. or d.c. input switch. Using the 'scope with the amplifier directly coupled has great advantages, since it shows the precise working conditions of amplifier and other stages. It is especially useful in transistor circuits where d.c. levels are low. In circuits where a small a.c. waveform has a large d.c. level, it would be inconvenient to have to use a low-sensitivity setting simply to overcome the d.c. bias, so the a.c. position of the input switch inserts a capacitor in circuit.

General Notes on Oscilloscope Use

A comprehensive guide to the use of an oscilloscope would fill a book of twice this size and could still leave much unsaid. In the elementary procedure for setting up an oscilloscope the recommendations of the maker should be followed; the notes here are on points of use which may be overlooked:

(1) When the type of signal which may be seen is unknown, as when delving for the first time into a circuit about which there is little information, always switch to maximum volts/cm, a.c. input, long sweep time and 'sync' rather than 'trigger'. When a trace is obtained, adjusting the Y sensitivity as needed, switch to a suitable timebase speed, and then to the 'trigger' setting. Final adjustments can then be made to the timebase speed and sensivity controls so that the amplitude of the trace fills about half of the screen and about two cycles of the waveform are seen. If the timebase has an 'auto' position, then the auto setting along with triggering can be used from the start.

(2) The operating conditions of many amplifier and pulse circuits are best checked using the d.c. input. The earth potential may be set by earthing the input of the 'scope and setting the trace line (timebase set to sync) to run along one of the graticule lines. The input can be then connected to the circuit being checked and d.c. levels read off as well as a.c. signal voltages. Remember that positive voltages produce upward deflections (even on Oriental scopes!). Be careful of polarity in circuits using p-n-p transistors.

(3) When a high speed timebase is in use, mains hum appears as a flickering of the whole waveform which can be mistaken for a high frequency signal. This can sometimes be removed by using a low value capacitance in series with the signal, providing that the effect of the capacitance is taken into account when making measurements.

(4) Remember that any oscilloscope has an input impedance, usually about 1 to 5 MΩ with 10 to 30pF in parallel. Signals coming from a high source impedance will be attenuated, and allowance must be made for this. If an external trigger is being used, the phase shifting effect of the input impedance may cause part of the trace to be missed.

(5) Except for all-transistor battery-operated oscilloscopes, the case and framework of an oscilloscope will be earthed. If one of the input terminals of the Y-amplifier is taken to the case and so earthed, it is then impossible to measure the waveform across a resistor when neither side is at earth potential. When the oscilloscope has a differential preamplifier there is no problem; and some oscilloscopes make provision for disconnecting the earthy side of the Y-amplifier from earth. Where neither of these facilities exists, some servicemen disconnect the oscilloscope mains earth lead.

This is a dangerous practice, and does not give reliable readings because of the large stray capacitance of the case. By far the best solution is to make up a simple differential preamplifier, battery operated, which can be plugged in when needed. Since the gain need only be unity (it is desirable, in fact, if the gain can be made exactly unity by the use of feedback) the bandwidth obtainable is wide enough for most purposes.

Servicing the Oscilloscope

In some respects, servicing a faulty oscilloscope is extremely easy, because the instrument displays its own faults. In other respects, notably in the calibration of timebase and Y-amplifier and correction of non-linearity, servicing is extremely difficult, as it may call for an oscilloscope of even better quality to examine the waveforms obtained; for such work the maker's handbook should always be consulted. Table 10.1 shows the checks which can be made without the use of elaborate apparatus.

Fault	Check first—	—then
No life	Mains input fuse	mains connections
Heaters on but otherwise dead	H.T. line fuse, then check for h.t. present	if no h.t., check rectifier circuits
Voltages present but no trace	shifts and brightness controls (spot may be biased or shifted off)	e.h.t. circuits
No timebase	connections to X plates, especially if plugs or switches used	timebase amplifier, then timebase
No Y deflection	connections to Y plates, especially if plugs or switches used	Y amplifier, pre-amp, input sockets
Dim trace or visible flyback	brilliance control setting	bright-up pulse circuits

Table 10.1 Faultfinding in the Oscilloscope

Oscilloscope Circuits

Fig. 10.2 shows details of the circuits of the timebase of a general-purpose oscilloscope, the Advance OS15. This timebase provides speeds of 100, 10 and 1mS/cm and 100, 10 and 1μS/cm with a typical measuring accuracy of $\pm 5\%$. In addition, the X-gain and variable time controls can be used to extend the timebase range to 1S/cm in the low speed direction and to 0·5μS/cm in the high speed

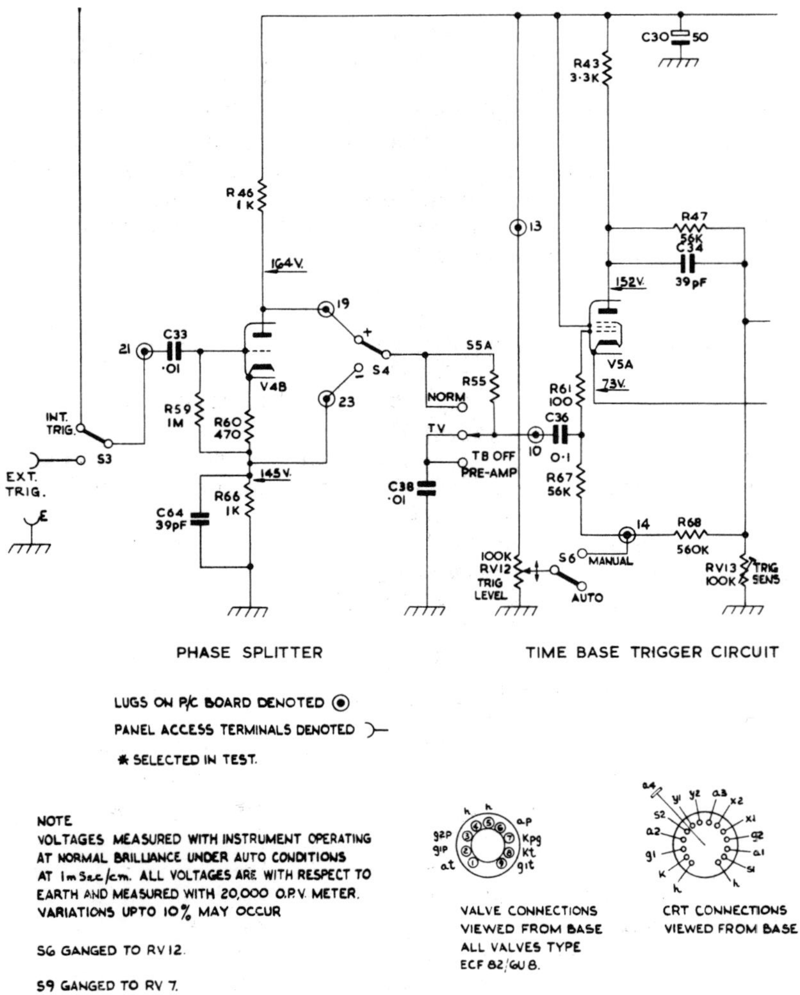

direction. The timebase can be triggered in several different ways, internally or from a separate triggering signal. 'Auto' trigger locks the timebase automatically to any signal of frequency within the range 50Hz to 1MHz, with positive- or negative-going pulses.

Using the trigger-level selector, rather than 'auto', signals of frequencies up to 2MHz can be triggered. Any signal which can give a deflection of 5mm on the screen, or an external trigger of 1V amplitude, will trigger the timebase. A timebase output of about

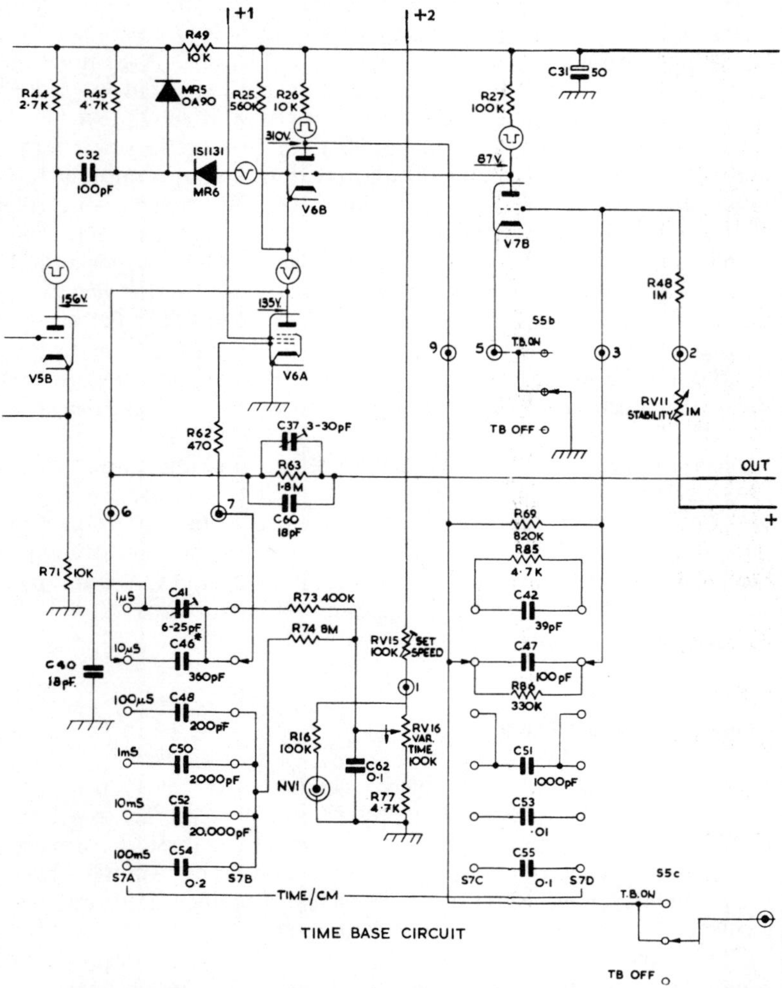

Fig. 10.2 Timebase of type OS15 oscilloscope. (*courtesy of Advance Electronics Ltd.*)

4·5V, negative-going, is provided at the front panel so that other equipment (such as wobbulators, see later) can be synchronised.

The timebase consists of a trigger circuit which is a cathode-coupled multivibrator driven by the input signal. In the 'manual' position of the switch *S*6, the voltage level at which the multivibrator will flip over is set by the 'trigger level' potentiometer; in the 'auto' position, the time constant *C*36, *R*67, *R*68 acts to prevent the flip-flop from operating at its natural rate and restricts it to about 50Hz. In this position, the multivib. will run at the speed of any signal fed in through *C*36 and will follow changes of speed.

The synchronising output is differentiated by *C*32, *R*45; positive portions are removed by *MR*5, and the signal is gated by *MR*6, which ensures that sync. signals cannot be transmitted while the grid of *V*6b is very much more negative than the anode side of the rectifier. *V*6b and *V*7b form a flip-flop with one direct coupling and one time-constant which is adjusted for different timebase speeds.

*V*6a is a Miller stage which is gated by the cascade connection of *V*6b in the anode. When the negative pulse from the trigger circuit cuts off *V*6b, the Miller valve *V*6a starts to generate the negative-going sweep, drawing current through *R*25. Eventually the falling voltage at the cathode of *V*6b causes that valve to switch on again, so making the anode voltage of *V*6a rise sharply and restore the original conditions. The switching action of *V*6b is rapid, as it is half of a flip-flop.

Note the large number of direct couplings, making it important, as is stressed in the maker's handbook, never to operate the 'scope with any valve removed, since this could cause the full line voltage to be applied to a grid of the next stage. Only the timebase of this 'scope has been examined here, since the techniques used in the Y-amplifier have already been covered in Chapter 5, as was part of the Y-amplifier of this 'scope.

Sinewave Measurements

Space does not permit an exhaustive account of the measuring uses of the oscilloscope, but a few representative measurements will be described, covering the range of measurements, as distinct from the observation of waveforms, most often required. Among these requirements in working with sinewaves, amplifier gain and the plotting of gain/frequency graphs rank highest.

The instruments needed in addition to the oscilloscope are a signal generator to cover the range of frequency required and an attenuator which can give a range of attenuation to an amount roughly equal to the gain of the amplifier and which can be used over the whole of the frequency band. The attenuator should preferably be calibrated in dB and used between known impedances. Many signal generators have such an attenuator built in.

To measure signal gain, the signal generator is wired through the attenuator to the input of the amplifier, using screened cables where necessary and taking care that the attenuator is terminated by the correct value of resistance if the scale is calibrated in dB. Set the attenuator to maximum attenuation, connect the oscilloscope input to the input of the attenuator and adjust the output of the signal generator at 1kHz to 1V, or other convenient value, as measured on the graticule of the 'scope.

Then disconnect the 'scope input from the attenuator and wire it to the output of the amplifier. With the amplifier gain setting at maximum, adjust the attenuator until the output of the amplifier is 1V, or whatever value was taken in the first reading. The reading on the attenuator in dB is now equal to the gain of the amplifier. This presumes that the attenuator was terminated by the same value of resistance when feeding into the 'scope as it was when feeding the amplifier. This type of reading avoids using the calibration of the 'scope and therefore can be used with even the simplest type of instrument.

If the gain is checked at several frequencies, the gain/frequency graph can be drawn. If the signal generator is known to be stable, giving out the same set amplitude of signal over the whole frequency range, then the check can be made quickly by noting the attenuator setting at different frequencies. Also in this way, the 3dB points can be very quickly found by setting the attenuator 3dB down from its setting at 1kHz and adjusting the frequency until the 'scope shows the set level again.

When gain/frequency graphs are to be drawn for a multistage r.f.-i.f. amplifier, the frequencies being handled are beyond the reach of the simpler type of 'scope, and the more common method of checking is to keep the modulation of the input constant and use a 'scope at the output of the detector. When a large number of adjustments have to be made, as for example to set up an amplifier to a specified bandwidth, then a wobbulator can be a most useful accessory to the oscilloscope.

The wobbulator provides an output which is swept in frequency in step with the timebase of the 'scope. The trace from the detector then represents the gain/frequency graph and can be calibrated by injecting frequencies from a conventional signal generator which provides marker points on the trace. The effect of adjustments on the gain and bandwidth can be seen at once as the adjustments are being made. To operate the wobbulator, a timebase voltage must be available from the 'scope.

Phase measurement cannot be made with such a degree of accuracy on a single beam 'scope. One technique is to trigger the 'scope through the external trigger input from the signal source and to display one cycle of input. Display first a wave of a frequency at which the phase shift might be expected to be small, and note at what point on the waveform the trace first appears; this will be only a very short time after the trigger pulse has started the timebase.

Now change the frequency to a value where a phase shift is to be expected. The timebase should trigger at about the same point on the waveform if the amplitude of input signal is the same, since the trigger is operated from the input, but the output waveform will have shifted in phase, and the first appearance of the trace will be at a different point on the waveform. The phase difference can then be estimated. This is not an accurate method, but phase shifts of the order of 45° or so are easily detected.

When a double-beam 'scope is available, the two traces can be used to display input and output waveforms simultaneously at about the same amplitude and the phase shift checked by superimposing and measuring, remembering that one waveform in a double-beam tube is inverted if the usual beam-splitting arrangement is used in the tube. The latter provision does not apply if two separate guns are used in the c.r.t.

Another method of phase measurement is noted later under the title 'Lissajou Figures'.

Pulse Measurements

The measurement of rise and fall times, sag, overshoot and undershoot of repetitive fairly wide pulses is straightforward with a good quality 'scope, but working with narrow pulses requires more care. If the pulse is narrow compared to the time between pulses, it will normally be impossible to see any detail of the pulse when the timebase has been triggered by the previous pulse because the duration of the timebase will be that of the time between the pulses, otherwise nothing will be seen at all.

If the pulse which has triggered the 'scope is itself viewed by setting to a high timebase speed, then only the trailing edge will be seen because of the time delay between the leading edge triggering the timebase and the timebase starting a linear sweep.

One way round this is to take the trigger pulse from the start of a delay line and examine the pulse by connecting the Y-input to the other end of the line. In this way, by the time the pulse arrives at the end of the line and enters the Y-amplifier, the timebase has already started. Though a delay line of this type was built into one early (and greatly revered) Solartron 'scope, it has not been widely used outside the realm of specially made instruments for radar work. For most oscilloscopes, an external wide-band delay line would have to be used.

Another way round the problem is available if the pulse is used for testing and is derived from a pulse generator which has a 'pre-pulse' trigger. This, as the name suggests, is a trigger pulse available from another output and slightly ahead of the main pulse. This enables the 'scope once again to be triggered slightly ahead of the arrival of the main pulse so that details of the leading edge of the main pulse can be seen.

Where the facility is not provided, it is possible to make up a flip-flop time delay, triggered by one pulse and generating a squarewave just slightly shorter than the time between pulses. Differentiating the output will give a trigger pulse from the trailing edge of this squarewave which occurs just ahead of the next pulse, and which can therefore be used as an oscilloscope trigger.

A few oscilloscopes, as mentioned earlier, can trigger a second timebase from a selected point on the main timebase, and then expand the second timebase to form the complete display of a small portion of the main trace, and this also can be used to view the start of a repetitive pulse. If the pulse is not repetitive, then only a storage oscilloscope, such as the Cawkell *Remscope*, together with a delay line can give a viewable display. Such oscilloscopes use a specially developed c.r.t. which stores a transient trace for long periods.

Where none of the facilities mentioned above are available, and repetitive narrow pulses have to be viewed, an expansion of the trace may be enough to do the job. The pulse after the trigger is viewed on a fairly long timebase, set to the time between pulses, and is centred in the display. The trace is then expanded, keeping the pulse central by adjustments to the X-shift, until the detail of the pulse can be seen.

Unfortunately, the expanded trace is usually uncalibrated, so that measurements cannot be made with any accuracy unless calibration pips from an oscillator, preferably crystal controlled, can be fed in and used as time markers. The feed of such pips can be to another trace, to the Z-mod, or, if there is no alternative, to the same Y-amplifier.

The use of calibration pips is often extremely useful in enabling measurements of high accuracy to be made with an oscilloscope which is itself not capable of high standards of measurements. In a similar way, a voltmeter of known accuracy can be coupled to the slider of the Y-shift potentiometer so that the shift of the trace can be calibrated to an accuracy as good as that of the meter, assuming that the Y-shift operates directly on the plates. If the Y-shift operates on the final paraphase amplifier input this method is not quite so useful.

Lissajou Figures

The use of Lissajou figures is a technique older than the oscilloscope but which does not appear to be so well known now as in earlier years. If a 'scope of high price and correspondingly good facilities is available, the use of Lissajou figures is seldom necessary, but when measurements have to be taken with a 'scope of simple design and limited facilities, then the use of Lissajou figures is excellent in measurements of frequency and phase.

To set up a Lissajou figure (Fig. 10.3), the timebase is put out of action and a sinewave signal applied to the X-amplifier or even to the X-plates if a signal of sufficient amplitude can be obtained. Another sinewave signal is then applied to the Y-amplifier. If the two sinewave signals are of amplitudes which cause equal amounts of beam deflection (this does not mean equal amplitudes, since the sensitivities of the two sets of plates differ) and one signal is using a different amplifier compared to the other, and if the signals are in phase, then the trace is a straight line at 45° to the horizontal.

If there is a phase difference between the signals the line becomes opened out into an ellipse, slowly opening as the phase difference increases until it is circular at a phase difference of 90°. The amount of the phase difference between these extremes can be obtained by measuring the major and minor axes of the ellipse (largest 'diameter'

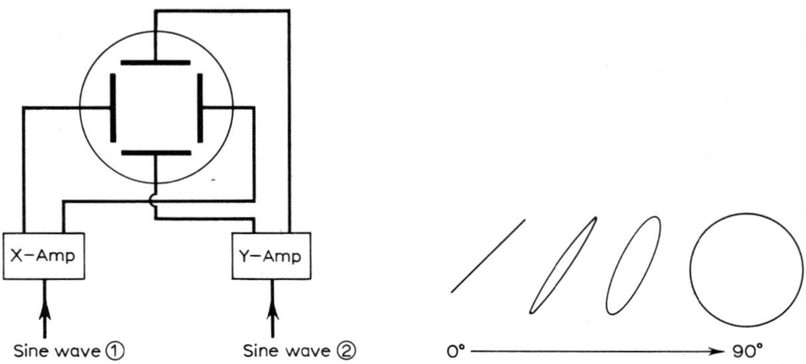

Fig. 10.3 Setting up a Lissajou figure.

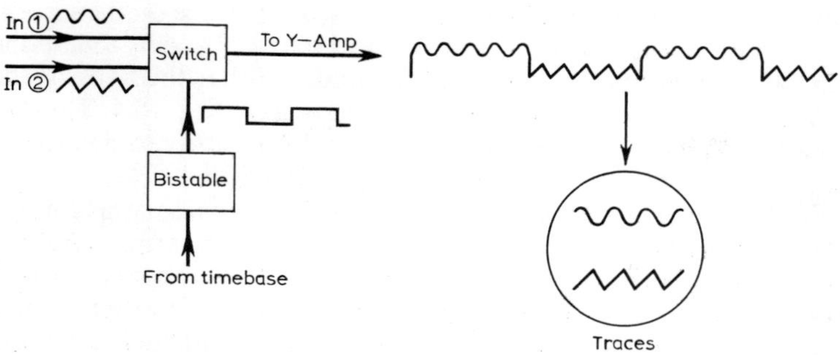

Fig. 10.4 Principle of beam-switch.

and shortest 'diameter') and taking the ratio; major/minor, which is the sine of the phase angle. This can be applied only when the signals produce equal amplitudes of deflection and the 90° shift trace is a circle.

When one frequency differs from the other, the trace changes shape at a speed depending on the difference between the frequencies, going from line through ellipse to circle and back twice in a cycle of frequency difference. This is an extremely useful way of comparing two frequencies which are close to one another—by taking the time for one set of changes of pattern, the number of Hz of frequency difference can be found; it is 1/time. The pattern-changes frequently present the interesting optical illusion that one is watching a 3-D rotating circle.

When the frequency difference is too great for the pattern 'revolutions' to be counted, sets of stable patterns can be obtained when the frequencies are related. When one frequency is twice that of the other, the pattern seen is a figure eight or some other two-looped pattern (what modern mathematics scholars call a 'topologically related pattern'). At higher frequency differences, different patterns are displayed whose number of loops give the ratio of the two frequencies. The patterns appear at rest only when the frequencies are exactly related.

Beam Switching

When an oscilloscope uses a tube with only one beam, there are almost inevitable occasions when the use of a second beam would be desirable. Double-beam 'scopes are, of course, available, but beam switching is an alternative method. A beam switch accepts two input signals and applies them in turn to the 'scope input together with a d.c. level which decides the position of the trace on the screen. In its simplest form, this consists of a bistable operated by the 'scope timebase and applied to a switching circuit so that the two signals are applied to the input on alternate sweeps, as in Fig. 10.4. The d.c. levels at the outputs of the switches then act as shifts to determine the position of the trace on the screen for each signal.

For low timebase speeds, this method is unsatisfactory, as the eye then sees the traces alternately instead of seeming superimposed and simultaneous, and an alternative method for slow traces is to speed up the bistable, triggering it from a multivibrator, so that several hundred transitions from high trace to low trace and back are made in each sweep. The low timebase speed then does not permit the eye to see the sharp rise and fall traces, and the appearance is of two continuous traces.

Sampling Oscilloscopes

Sampling oscilloscopes are used for examining repetitive waveforms at frequencies too high to be handled by conventional oscilloscopes; sampling adaptors can also be used with conventional oscilloscopes. A sampler contains a switch circuit which closes for a very short period of time during each waveform (or every second or third etc.), so switching a portion of the waveform to the output of the switch.

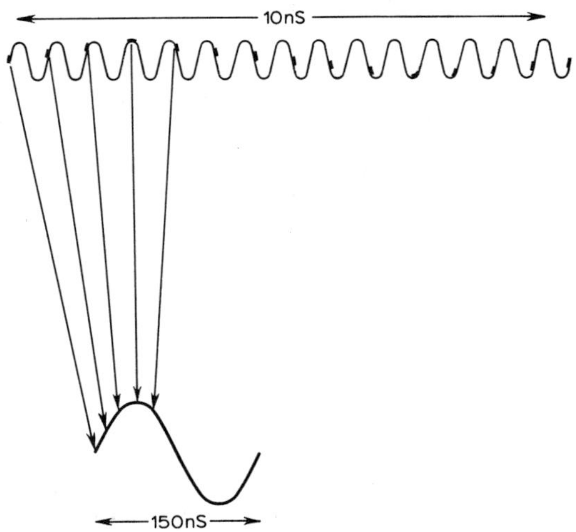

Fig. 10.5 Sampling. The small portions sampled on each waveform are brought together on a slower timebase. In practice, many more samples would be taken and a longer timebase used.

The switching is arranged to be triggered slightly later in the waveform each time, an amount equal to the time of the portion switched, so that over a complete cycle of triggering, which may be very long compared to a cycle of the waveform, a composite picture of the wave is built up, as in Fig. 10.5. The timebase of the 'scope is then set to sync on this slower composite waveform.

There is, of course, a frequency limit to the process, because the switch must be capable of being opened and closed for a time which is brief compared to a wave time, so that the performance of the switch must be excellent. This, however, is the only part of the circuit which has to operate at very high speed, and this makes the sampling method very useful for waveforms which cannot be handled by amplifiers but which can be switched, even if only by devices beyond the reach of conventional laboratory apparatus. For small amplitude signals, tunnel diodes have been found capable of carrying out switching at speeds sufficient to run sampling 'scopes on waveforms of several GHz.

INDEX

Index